Analytical Assessment of e-Cigarettes

Emerging Issues in Analytical Chemistry

Series Editor
Brian F. Thomas

AMSTERDAM • BOSTON • HEIDELBERG • LONDON
NEW YORK • OXFORD • PARIS • SAN DIEGO
SAN FRANCISCO • SINGAPORE • SYDNEY • TOKYO

Analytical Assessment of e-Cigarettes

From Contents to Chemical and Particle Exposure Profiles

Konstantinos E. Farsalinos
Onassis Cardiac Surgery Center, Athens, Greece and University of Patras, Patras, Greece

I. Gene Gillman
Enthalpy Analytical, Inc., Durham, NC, United States

Stephen S. Hecht
University of Minnesota, Minneapolis, MN, United States

Riccardo Polosa
University of Catania, Catania, Italy

Jonathan Thornburg
RTI International, Research Triangle Park, NC, United States

ELSEVIER

AMSTERDAM • BOSTON • HEIDELBERG • LONDON
NEW YORK • OXFORD • PARIS • SAN DIEGO
SAN FRANCISCO • SINGAPORE • SYDNEY • TOKYO

Elsevier
Radarweg 29, PO Box 211, 1000 AE Amsterdam, Netherlands
The Boulevard, Langford Lane, Kidlington, Oxford OX5 1GB, United Kingdom
50 Hampshire Street, 5th Floor, Cambridge, MA 02139, United States

Published in cooperation with RTI Press at RTI International, an independent, nonprofit research institute
that provides research, development, and technical services to government and commercial clients worldwide
(www.rti.org). RTI Press is RTI's open-access, peer-reviewed publishing channel. RTI International is a trade
name of Research Triangle Institute.

British Library Cataloguing-in-Publication Data
A catalogue record for this book is available from the British Library

Library of Congress Cataloging-in-Publication Data
A catalog record for this book is available from the Library of Congress

ISBN: 978-0-12-811241-0

Publisher: John Fedor
Acquisition Editor: Kathryn Morrissey
Editorial Project Manager: Amy Clark
Production Project Manager: Paul Prasad Chandramohan
Cover Designer: Mathew Limbert

Typeset by MPS Limited, Chennai, India

To my parents, Efthymios and Penelopi, for their devotion to family values and to the intellectual and social education of myself and my siblings.

K. E. Farsalinos

CONTENTS

LIST OF CONTRIBUTORS

N.L. Benowitz
University of Calilfornia, San Francisco, San Francisco, CA,
United States

K.E. Farsalinos
Onassis Cardiac Surgery Center, Athens, Greece and University of Patras,
Patras, Greece

I. Gene Gillman
Enthalpy Analytical, Inc., Durham, NC, United States

S.S. Hecht
University of Minnesota, Minneapolis, MN, United States

R. Polosa
University of Catania, Catania, Italy

J. Thornburg
RTI International, Research Triangle Park, NC, United States

FOREWORD

Cigarette smoking has had a devastating effect on public health worldwide over the past 100 years and will continue to do so throughout the current century unless there is a substantial reduction in the prevalence of smoking. Compulsive smoking is driven by addiction to nicotine, but most of the harm from smoking is caused by exposure to tobacco combustion products. For many years, tobacco researchers and policy experts have entertained the idea that a clean source of nicotine that could be inhaled and provide similar rewarding effects as a cigarette might entice smokers away from cigarette smoking and lead either to quitting smoking or to long-term nicotine use without incurring the harm from tobacco combustion toxicants.

e-Cigarettes are nicotine delivery devices that deliver nicotine without combusting tobacco. These are battery-powered devices that heat a liquid composed of propylene glycol and/or vegetable glycerin, nicotine, and flavoring to form a vapor which rapidly aerosolizes and is inhaled like cigarette smoke. e-Cigarettes could be beneficial to public health if they help smokers quit smoking and possibly (at least for some health effects) reduce harm for those who smoke fewer cigarettes while using e-Cigarettes. On the other hand, there are several concerns about adverse effects of e-Cigarette use on a population level, including attracting youth and serving as a gateway to nicotine addiction and cigarette smoking, dual use with cigarettes resulting in lower rates of quitting smoking, renormalizing nicotine use and undermining smoke-free air legislation, and/or diverting smokers from proven smoking cessation treatment sessions.

One of the determinants of the net effect of e-Cigarettes on public health is the benefit versus harms, including the direct toxicity of e-Cigarette use. Assessing the toxicity of e-Cigarettes requires an understanding of the design and variability in device components and constituents of e-liquids and aerosols, both chemicals and particulates. Biomarkers of exposure to e-Cigarette toxicants in people are critical for extrapolating machine-tested e-Cigarette emission findings to actual human exposures.

This volume brings together a number of expert e-Cigarette researchers to explore the contents, chemical and particulate emissions, and human exposure to e-Cigarette constituents, as well as regulatory issues related to e-Cigarettes. The authors are Dr. Konstantinos Farsalinos, a cardiologist who has conducted research on e-Cigarette use epidemiology, the nature of e-Cigarette emissions, and cardiovascular effects of e-Cigarettes; Dr. I. Gene Gillman, an analytical chemist who has measured constituents of e-Cigarette aerosols; Dr. Jonathan Thornburg, an aerosol physicist who has studied e-Cigarette aerosol composition; Dr. Stephen Hecht, a toxicologist who has pioneered research on biomarkers of exposure to tobacco constituents; and Dr. Riccardo Polosa, a pulmonary physician who has conducted clinical trials of e-Cigarettes for smoking cessation and reduction, and has studied pulmonary effects of e-Cigarette use. These researchers present a highly informative review of the current state of understanding of the contents and toxicant properties of e-Cigarettes as of 2016. The material in this book will be of great interest to researchers and regulators addressing the health consequences of e-Cigarette use.

N.L. Benowitz
Professor of Medicine and Bioengineering & Therapeutic Sciences
Chief, Division of Clinical Pharmacology
University of California, San Francisco
San Francisco, CA, United States
October 2016

CHAPTER 1

Introduction to e-Cigarettes

K.E. Farsalinos

INTRODUCTION

e-Cigarettes were recently invented and developed as an alternative-to-smoking method of nicotine intake. They are electronic devices with three main parts: a battery, an atomizer composed of a wick and metal coil, and a liquid ("e-liquid") stored inside the atomizer. The function is to aerosolize the liquid, producing a visible aerosol which the user inhales. This is achieved by heating the metal coil inside the atomizer by an electrical current from the battery. e-Cigarettes are commonly called electronic nicotine-delivery devices (ENDS). However, because they can be used with non-nicotine e-liquids, the term ENDS is inaccurate and does not represent the whole spectrum of e-Cigarette devices and use patterns.

INVENTION AND EVOLUTION OF e-CIGARETTES

e-Cigarettes were invented by Hon Lik, a pharmacist from China. The initial patent was filed in 2004 in China and in 2005 in the United States Patent Office (application number 10/587,707).[1] The patent was published in 2007. The principle of e-Cigarette function, the evaporation of liquid and delivery of aerosol to the user, can be tracked back to patents published decades earlier. In 1930, a patent was published by the US Patent Office describing an electric vaporizer "for holding medicinal compounds which are electrically or otherwise heated to produce vapors for inhalation."[2] Another patent in 1934 described a therapeutic apparatus that was "adapted for transforming volatile liquid medicaments into vapors or into mists of exceedingly fine particles."[3] A similar patent was published in 1936.[4] These cases referred to vaporization for therapeutic applications. However, a patent filed by Herbert A. Gilbert and published in 1965 was titled "Smokeless

Analytical Assessment of e-Cigarettes. DOI: http://dx.doi.org/10.1016/B978-0-12-811241-0.00001-2

non-tobacco cigarette" and described a battery-operated device "to provide a safe and harmless means for and method of smoking by replacing burning tobacco and paper with heated, moist, flavored air."[5]

The term e-Cigarette includes a very diverse line of products, with different design, functional, and performance characteristics. Although there is still no consensus on terminology, the products available on the market are mainly of three types (Fig. 1.1).[6]

1. First-generation ("cigalike") devices have similar size, shape, and appearance as tobacco cigarettes. They consist of a small lithium battery and a cartomizer. The battery can be either disposable

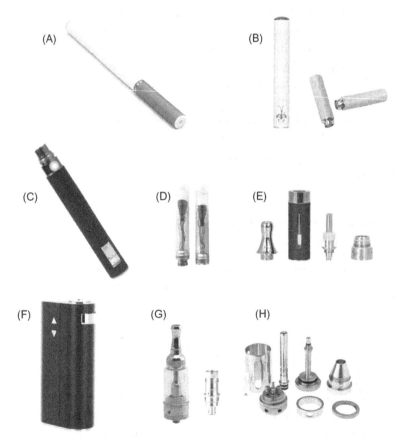

Figure 1.1 Types of e-Cigarettes. First generation: (A) disposable cigalike; (B) rechargeable cigalike and replacement prefilled cartomizers. Second generation: (C) eGo-type battery; (D) tank-type atomizers; (E) tank-type atomizer with replacement coil and wick head. Third generation: (F) box-shape variable-wattage battery device; (G) tank-type atomizer with replacement head; (H) rebuildable atomizer.

(discarded after being discharged) or rechargeable. The cartomizer is a specific type of atomizer, consisting of a sponge-like polyfil (polyester fiber) material which is soaked with the liquid. There is no tank where the liquid is stored. Usually, cartomizers are prefilled with liquid; however, empty cartomizers are also available, which can be refilled with any liquid of choice. These were the first e-Cigarettes released to the market.

2. Second-generation devices consist of a rechargeable lithium battery of larger size and cylindrical shape, resembling a large pen. These are called eGo-type batteries. The atomizer is refillable and has a tank design, with a storage space for the liquid and a transparent window so that the user can see the level of the liquid. Initially, the whole atomizer had to be discarded after several milliliters of liquid consumption, but in the last few years they have been available with removable heads so that the resistance and wick parts are replaced and the body retained. This significantly reduced the cost of use and resulted in the development of more advanced products.

3. Third-generation devices, called "mods" or "advanced personal vaporizers" (APVs), consist of a large-capacity lithium battery with an integrated circuit that allows the user ("vaper") to adjust the energy (wattage) delivered to the atomizer. They usually have either a cylindrical or a box-like shape. They can be combined with either second-generation atomizers or rebuildable atomizers, where the consumers can prepare a custom setup of resistance and wick. Most of these atomizers have a tank-type design, but there is a distinct group of "dripper" atomizers which have no storage space; the user introduces the liquid from the mouthpiece at regular intervals to keep the wicking material wet.

Another proposed classification is open versus closed systems, referring to the ability or not to refill the atomizer with liquid. However, closed systems can be modified and refilled. Furthermore, this classification is not indicative of different functional and performance characteristics and is of limited value for the understanding of consumers about the potential of the devices.

Since the initial introduction of e-Cigarettes to the market, new products have evolved rapidly. Cigalike devices, resembling the tobacco cigarette in shape, form, weight, and function, have low aerosol volume production[7] and low nicotine delivery potential.[8,9] This was evident from surveys of dedicated users, which showed preference for newer generation e-Cigarettes.[10–12] The newer devices produce more sensory satisfaction[11] and more nicotine delivery and absorption,[9] very close to the delivery rate and level of tobacco cigarettes.[13] There has also been progress in construction materials, especially for atomizers. Current atomizers use pyrex glass and stainless steel instead of plastics and other metals, while the wicking material is cotton instead of silica[14]; still, no research has determined whether these developments are accompanied by fewer harmful emissions in the aerosol. A wide variety of liquids are available, with thousands of flavors[15] and different nicotine content (including non-nicotine liquids). The main ingredients of the liquids are the humectants propylene glycol and glycerol, as well as flavoring compounds. The latter are either natural extracts or synthetically produced substances and are, in most cases, approved and generally recognized as safe for ingestion.

TOBACCO HARM REDUCTION AND e-CIGARETTES

Harm reduction is the strategy, policy, and philosophy of reducing risk and thus the morbidity and mortality associated with an action or condition. Widely known examples are needle and syringe exchange programs and opioid substitution therapy for intravenous drug users to reduce the risk of blood-borne infectious diseases such as hepatitis and HIV.[16,17] These approaches have been actively endorsed by authorities such as the World Health Organization[18] and have been integrated into the legislation of several countries.[19] They have proven to be cost-effective and to reduce risk and improve quality of life.[20] Beyond that, the philosophy of harm reduction is strongly endorsed in everyday social activities, with characteristic examples being the use of seat belts in cars, helmets for motorcycles, and condoms in sexual activities.

Tobacco harm reduction seeks to decrease the net damage to health associated with the use of combustible tobacco products. It provides alternative sources of nicotine to smokers who are unable or unwilling to quit tobacco and nicotine entirely. It is based on the concept that "smokers smoke for nicotine but die from tar," expressed by British tobacco addiction researcher Michael A.H. Russell,[21] referring to combustion products and toxins other than nicotine which are present in smoke. Although nicotine itself may not be absolutely harmless, several studies evaluating the effects of non-combustible nicotine products have shown that it is highly unlikely to contribute significantly to smoking-related cancer and cardiovascular disease.[22−26] Russell proposed a harm reduction approach in 1974.[27] He realized the high dependence-producing potency and the broad appeal of the effects of nicotine on smokers and recognized that "the goal of abstinence and the abolition of all smoking are unrealistic and doomed to fail." A few years later, smokeless tobacco products were proposed as harm reduction agents.[28,29] Since then, several studies have shown that their use can reduce some smoking-related diseases. The most characteristic case is in Sweden, where tobacco use prevalence in males is high but is mostly Scandinavian snus use rather than smoking. As a result, Sweden has the lowest death rates from cancer and cardiovascular disease among European Union countries.[30] Recently, major health organizations such as the US Food and Drug Administration and the UK Medicines and Healthcare Products Regulatory Agency have accepted long-term nicotine therapy for harm reduction in smokers.[31−33]

Smoking dependence is not solely attributed to nicotine. The sensorimotor aspects and rituals of the smoking act have an important role.[34,35] e-Cigarettes are the only products in the current arsenal that replicate the rituals along with nicotine delivery. As such, they produce harm reduction in three ways: allowing smokers to quit, helping former smokers avoid relapse, and preventing non-smokers from initiating smoking. From a public health perspective, e-Cigarettes should be only used as a harm reduction product. However, they may also be used as a new habit by never-smokers or people not intending to smoke tobacco cigarettes. Population studies have shown that regular use of e-Cigarettes occurs mostly among smokers and former smokers, supporting the argument that they are used for harm reduction. However, experimentation is observed at increasing rates among non-smokers, and this should be continuously monitored to assess if they become regular users, consume nicotine-containing e-Cigarettes, or transition to smoking. There has been an exponential increase in e-Cigarette awareness and use over the last few years.[36,37] As expected, this has attracted the interest of researchers, public health, governments, and regulators.

RESEARCH ON e-CIGARETTES

Intense research on all aspects related to e-Cigarettes, including chemistry, toxicology, clinical effects, and population effects, is ongoing.

e-Cigarettes are complex products. Liquids have many ingredients, particularly flavoring compounds. Although no chemical was specifically synthesized or developed to be used in e-Cigarettes, and almost all compounds used are approved for human consumption, their safety has been mostly assessed for ingestion. With e-Cigarettes, the liquid is subject to heating and evaporation, resulting in the emission of an aerosol. The aerosol is inhaled instead of ingested, which leads to direct lung exposure and fast absorption directly into the arterial circulation, bypassing the first-pass metabolism in the liver. The atomizers, where the liquid is stored, have several metal and plastic components, and there may be interaction between these and the liquids, resulting in emission of harmful substances. The heating process and the aerosol yield are highly dependent on the design and structure of the atomizer and the energy delivered from the battery. The huge variability of e-Cigarette devices and liquids makes evaluation of the aerosol composition complex.

Chemical evaluation is a vital step in the assessment of many consumer products. It is important for informing consumers about the potential benefits and risks of exposure and informing regulators to make appropriate decisions to ensure quality and safety. In the case of e-Cigarettes, the chemistry of the emitted aerosol is critical; this is what the user inhales. Obviously, there is overlap between liquid composition and aerosol emissions, but the heating process can result in the de novo formation of chemicals not present in the liquid formulation. The following chapters will discuss analytical aspects of the aerosol, assessment of biomarkers of exposure, regulatory decisions that could ensure the quality of products, and risk assessment, especially for smokers who make the partial or complete switch to e-Cigarettes. Research is continuously evolving and progressing, and the chapters will address current knowledge and future prospects.

REFERENCES

1. United States Patent Application Publication. Electronic atomization cigarette. Publication No: US 2007/0267031 A1; November 22, 2007.

2. United States Patent Office. Electric vaporizer. Publication No: 1,775,947; September 16, 1930.

3. United States Patent Office. Therapeutic apparatus. Publication No: 1,968,509; July 31, 1934.

4. United States Patent Office. Vaporizing unit for therapeutic apparatus. Publication No: 2,057,353; October 13, 1936.

5. United States Patent Office. Smokeless non-tobacco cigarette. Publication No: 3,200,819; August 17, 1965.

6. Farsalinos KE, Polosa R. Safety evaluation and risk assessment of electronic cigarettes as tobacco cigarette substitutes: a systematic review. *Ther Adv Drug Saf* 2014;**5**:67−86.

7. Farsalinos KE, Yannovits N, Sarri T, Voudris V, Poulas K. Protocol proposal for, and evaluation of, consistency in nicotine delivery from the liquid to the aerosol of electronic cigarettes atomizers: regulatory implications. *Addiction* 2016;**111**:1069−76.

8. Nides MA, Leischow SJ, Bhatter M, Simmons M. Nicotine blood levels and short-term smoking reduction with an electronic nicotine delivery system. *Am J Health Behav* 2014;**38**:265−74.

9. Farsalinos KE, Spyrou A, Tsimopoulou K, Stefopoulos C, Romagna G, Voudris V. Nicotine absorption from electronic cigarette use: comparison between first and new-generation devices. *Sci Rep* 2014;**4**:4133.

10. Dawkins L, Turner J, Roberts A, Soar K. Vaping" profiles and preferences: an online survey of electronic cigarette users. *Addiction* 2013;**108**:1115−25.

11. Etter JF. Throat hit in users of the electronic cigarette: an exploratory study. *Psychol Addict Behav* 2016;**30**:93−100.

12. Farsalinos KE, Romagna G, Tsiapras D, Kyrzopoulos S, Voudris V. Characteristics, perceived side effects and benefits of electronic cigarette use: a worldwide survey of more than 19,000 consumers. *Int J Environ Res Public Health* 2014;**11**:4356−73.

13. Lopez AA, Hiler MM, Soule EK, Ramôa CP, Karaoghlanian NV, Lipato T, et al. Effects of electronic cigarette liquid nicotine concentration on plasma nicotine and puff topography in tobacco cigarette smokers: a preliminary report. *Nicotine Tob Res* 2016;**18**:720−3.

14. Farsalinos KE, Voudris V, Poulas K. Are metals emitted from electronic cigarettes a reason for health concern? A risk-assessment analysis of currently available literature. *Int J Environ Res Public Health* 2015;**12**:5215−32.

15. Zhu SH, Sun JY, Bonnevie E, Cummins SE, Gamst A, Yin L, et al. Four hundred and sixty brands of e-cigarettes and counting: implications for product regulation. *Tob Control* 2014;**23** (Suppl. 3):iii3−9.

16. Marshall BDL, Wood E. Toward a comprehensive approach to HIV prevention for people who use drugs. *J Acquir Immune Defic Syndr* 2010;**55**(Suppl 1):S23−26.

17. Beyrer C, Malinowska-Sempruch K, Kamarulzaman A, Kazatchkine M, Sidibe M, Strathdee SA. Time to act: a call for comprehensive responses to HIV in people who use drugs. *Lancet* 2010;**376**:551−63.

18. World Health Organization, United Nations Office on Drugs and Crime, Joint United Nations Program on HIV/AIDS. HO, UNODC, UNAIDS technical guide for countries to set targets for universal access to HIV prevention, treatment and care for injecting drug users: 2012 revision.

19. Harm Reduction International. *Global state of harm reduction 2014*. Available from: https://www.hri.global/contents/1524.

20. Wilson DP, Donald B, Shattock AJ, Wilson D, Fraser-Hurt N. The cost-effectiveness of harm reduction. *Int J Drug Policy* 2015;**26**(Suppl. 1):S5−11.

21. Russell MA. Low-tar medium-nicotine cigarettes: a new approach to safer smoking. *Br Med J* 1976;**1**:1430−3.

22. Luo J, Ye W, Zendehdel K, Adami J, Adami HO, Boffetta P, et al. Oral use of Swedish moist snuff (snus) and risk for cancer of the mouth, lung, and pancreas in male construction workers: a retrospective cohort study. *Lancet* 2007;**369**:2015−20.

23. Lee PN, Hamling J. Systematic review of the relation between smokeless tobacco and cancer in Europe and North America. *BMC Med* 2009;**7**:36.

24. Huhtasaari F, Lundberg V, Eliasson M, Janlert U, Asplund K. Smokeless tobacco as a possible risk factor for myocardial infarction: a population-based study in middle-aged men. *J Am Coll Cardiol* 1999;**34**:1784–90.

25. Hansson J, Galanti MR, Hergens MP, Fredlund P, Ahlbom A, Alfredsson L, et al. Use of snus and acute myocardial infarction: pooled analysis of eight prospective observational studies. *Eur J Epidemiol* 2012;**27**:771–9.

26. Hansson J, Galanti MR, Hergens MP, Fredlund P, Ahlbom A, Alfredsson L, et al. Snus (Swedish smokeless tobacco) use and risk of stroke: pooled analyses of incidence and survival. *J Intern Med* 2014;**276**:87–95.

27. Russell MA. Realistic goals for smoking and health: a case for safer smoking. *Lancet* 1974;**1**:254–8.

28. Russell MAH, Jarvis MJ, Feyerabend C. A new age for snuff? *Lancet* 1980;**1**:474–5.

29. Kirkland LR. The nonsmoking uses of tobacco. *N Engl J Med* 1980;**303**:165.

30. Ramström L, Wikmans T. Mortality attributable to tobacco among men in Sweden and other European countries: an analysis of data in a WHO report. *Tob Induc Dis* 2014;**12**:14.

31. Medicines and Healthcare Products Regulatory Agency. Nicotine replacement therapy (NRT): new extended indication and consultation. Available from: http://webarchive.nationalarchives. gov.uk/20141205150130/http://mhra.gov.uk/safetyinformation/safetywarningsalertsandrecalls/ safetywarningsandmessagesformedicines/con068572; 2010.

32. National Institute for Health and Care Excellence. *Tobacco-harm-reduction approaches to smoking: guidance*. Available from: http://www.nice.org.uk/nicemedia/live/14178/63996/63996. pdf; 2013.

33. Food and Drug Administration (FDA). *Consumer health information. Nicotine replacement therapy labels may change*. Available from: http://www.fda.gov/downloads/ForConsumers/ ConsumerUpdates/UCM346012.pdf; April 2013.

34. Rose J, Levin E. Inter-relationships between conditioned and primary reinforcement in the maintenance of cigarette smoking. *Br J Addict* 1991;**86**:605–9.

35. Hajek P, Jarvis M, Belcher M, Sutherland G, Feyerabend C. Effect of smoke-free cigarettes on 24 h cigarette withdrawal: a double-blind placebo-controlled study. *Psychopharmacology (Berl)* 1989;**97**:99–102.

36. Dockrell M, Morrison R, Bauld L, McNeill A. e-Cigarettes: prevalence and attitudes in Great Britain. *Nicotine Tob Res* 2013;**15**:1737–44.

37. Adkison SE, O'Connor RJ, Bansal-Travers M, Hyland A, Borland R, Yong HH, et al. Electronic nicotine delivery systems: international tobacco control four-country survey. *Am J Prev Med* 2013;**44**:207–15.

CHAPTER 2

Analytical Testing of e-Cigarette Aerosol

I. Gene Gillman

INTRODUCTION

In the preparation of this chapter, the author read approximately 150 peer-reviewed publications and a large number of recent conference proceedings. Two things became apparent: (1) electronic nicotine delivery systems (ENDS) are evolving rapidly as manufacturers innovate, and the peer-reviewed literature does not adequately reflect products on the market. (2) Published findings of analytical testing and analysis of ENDS tend to show thermal decomposition products at either very low or very high levels, with little agreement in the middle. To address the first issue, recent conference proceedings are cited to fill gaps in the literature. The second issue is addressed only when required to highlight differences in testing methodologies. The objective is primarily to explain the results and implications of ENDS aerosol testing. Definitive advice on the choice and appropriateness of particular test methods must await further developments.

All current ENDS deliver nicotine in the form of a heated aerosol commonly comprised of propylene glycol (PG), glycerin (GLY), water, and flavoring. The modern ENDS arose from a 2003 invention by Chinese pharmacist Lik Hon,[1] but the concept of delivering nicotine in a heated PG or GLY aerosol has been around since at least 1960.[2] Technical challenges, including battery technology, delayed commercialization until the 21st century. ENDS are increasingly popular, with millions of users in the United States and Europe,[3–5] and are often a replacement for combustible cigarettes.[6] The aerosol is called a vapor, and usage is called vaping. This is inaccurate, because "vapor" refers to a gaseous state, while the aerosol from ENDS is a complex suspension of fine particles, not pure gas phase vapor compounds. The ENDS aerosol is a complex mixture containing semiliquid particulate matter along with gas phase compounds. Unlike combustible cigarette

Analytical Assessment of e-Cigarettes. DOI: http://dx.doi.org/10.1016/B978-0-12-811241-0.00002-4

smoke, ENDS aerosol does not contain true particulate matter. It is produced by atomization of the liquid and not through combustion of carbon-based material, as in smoking a combustible cigarette. The aerosol is typically generated by heating the liquid with a small gauge metal wire in contact with a heat-resistant material containing the liquid.

Two main types of atomizers are currently used. Devices with a separate wick to transfer the liquid from a reservoir to the heating coil are called clearomizers or atomizers. Those that integrate the heating coil into the liquid chamber are called cartomizers. In all devices, aerosol is delivered for inhalation into the user's respiratory tract.[7] Most of the aerosol is retained by the respiratory tract, and any not retained is exhaled into the environment.[8] ENDS come in a variety of shapes and sizes, from small devices that resemble a conventional combustible cigarette and are called cigalikes, to much larger devices called personal vaporizers, mods, or advanced devices. The larger devices visually have more in common with other electronic devices than with combustible cigarettes. ENDS have been commercialized in various forms and designs but all have similar physical and operational characteristics: a battery, a liquid reservoir, and an atomizer. The size of the product and amount of aerosol that can be delivered without recharging the battery are driven by the capacity of the battery. Small cigalikes can deliver the contents of a single, prefilled cartridge of around 1 mL of liquid on a single charge. Larger devices can deliver the contents of 5–40 mL before recharging. Table 2.1 gives the characteristics of common ENDS, including the chronological generation of each product, which is often cited in the literature.

The term "mods" has several meanings in the peer-reviewed literature and by end users. The original meaning sprang from the end user modifications to improve performance, such as the once common practice of constructing devices using flashlight bodies and large capacity batteries. It has also been used incorrectly to describe any variable voltage or variable power ENDS. It now means atomizers and tanks that can be modified by the end user and will be so used here.

A cigalike operates when the user draws air through it, which triggers an airflow sensor and activates a battery that powers an atomizer to produce an aerosol from liquid in the cartomizer (Fig. 2.1). Advanced devices function in much the same way except that power is applied to the atomizer coil only when a button is depressed by the user.

Generation	Types of Devices	Construction	Battery (mAh)	Comments
First	Disposable cigalike	Single unit	90–200	Prefilled
			Fixed voltage	
First	Rechargeable cigalike	Reusable battery	90–200	Prefilled
			Fixed voltage	
First	Three piece tank system	Separate battery, atomizer, and tank	90–200	Tank based (obsolete)
			Fixed voltage	
Second	Refillable ENDS with variable voltage	Separate battery and tank	300–1100	Tank based
			Variable voltage	
Second (mods)	Refillable ENDS with unregulated battery power	Separate battery and tank	300–1100	Tank based
			Voltage and wattage based on battery output	
Third	Refillable ENDS with variable voltage and power	Separate battery and tank	300–1100	Tank based
			Variable voltage, wattage	
Fourth	Refillable ENDS with variable voltage, power, and temperature	Separate battery and tank	>1000	Tank based
			Variable voltage, wattage, and temperature	

Table 2.1 Characteristics of Common ENDS

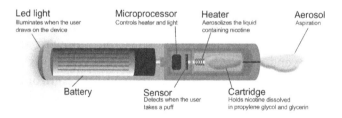

Figure 2.1 Diagram of a cigalike ENDS.

Current cigalikes are integrated cartomizer devices that contain the atomizer and liquid in one unit. The liquid reservoir is comprised of an adsorbent material that transfers the liquid to a wick wrapped by the heating coil. The wicking material is usually silica based but may be other materials. The adsorbent material in the cartomizer also helps to contain the liquid and prevent leaks (Fig. 2.2).

Refillable ENDS, unlike cigalike devices, allow the user to select and add liquid. These products typically contain a rechargeable power supply unit with a separate, refillable tank-based atomizer.

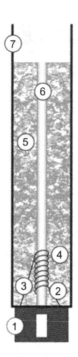

Figure 2.2 Diagram of cartomizer used in cigalike ENDS. (1) Battery connecter, (2 and 3) nonresistance wires, (4) resistance heating wire, (5) liquid reservoir with absorbent material, (6) aerosol exit, and (7) mouth end.

The power supply unit and tank-based atomizer are available from many suppliers and are usually interchangeable, allowing the user to select different vendors for the two components. Fig. 2.3 is a diagram of a rechargeable, variable power, temperature-controlled power supply unit.

One type of refillable tank-based atomizer is the clearomizer. It is usually cylindrical in shape with a clear plastic or glass tank and a separate wick, usually silica, that transfers the liquid from the tank to the heating coil. A clear tank allows the user to see the level of liquid. The wick may be placed at either the top or bottom of the tank (Fig. 2.4, Tanks A and B). Top coil tanks require longer wicks and thus have poor liquid-wicking rates, are more prone to dry puffing, and are limited in the maximum power that can be applied.

Another type of refillable is the cartomizer. It too is usually cylindrical, with a clear plastic or glass tank containing a replaceable atomizer coil. The liquid is transferred directly from the tank to the cartomizer containing the coil (Figure 2.4, Tank C). The coil may be

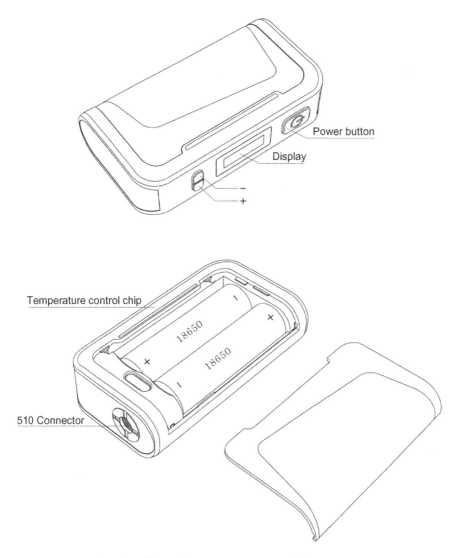

Figure 2.3 Diagram of a rechargeable, variable power, temperature-controlled ENDS power supply unit.

oriented horizontally or vertically. The wicking material can be made from a variety of materials, including silica, polyfill, cotton, or other heat-resistant absorbent material. Since the cartomizer must be in contact with the liquid, it is usually placed at the bottom of the tank. Because of the use of very short wicks and direct contact with the liquid, cartomizer tanks have good liquid-wicking rates, are less prone to dry puffing, and are available in a wide variety of power ranges.

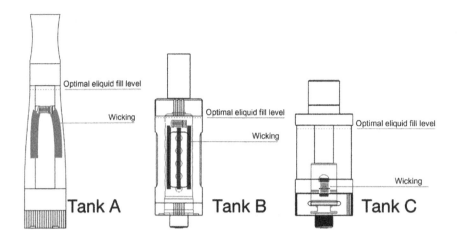

Figure 2.4 Examples of two top coil clearomizers (Tanks A and B) and a bottom coil cartomizer (Tank C). Wicking material (not shown) extends into the atomizer.

Refillable atomizers that allow for user customization are commonly called "mods." They may include rebuildable atomizers (RBAs), which allow the user to select and assemble the wick and coil materials instead of using commercially produced atomizers. RBAs are of two main categories: rebuildable tank atomizers (RTAs) and rebuildable dripping atomizers (RDAs). RTAs are similar in design to clearomizers in that a wick transfers the liquid from the tank to the coil. They are very customizable and offer many wicking and coil materials. RDAs are devices where the liquid is added or "dripped" by the user directly onto the coil and wick. These devices do not contain a tank. Because the amount of liquid is small, they are prone to overheat during active puffing due to a lack of liquid supply to the coil. Very few studies have been conducted on RDAs, but the limited research indicates that they may produce significant thermal decomposition products during normal use.[9] Given the range of configurations that can be made by the consumer and thus their nonstandard design, mods will not be covered in detail here.

Another design feature of ENDS is temperature regulation of the atomizer coil. Nontemperature-regulated devices use Nichrome or Kanthal wires because their physical properties, including resistance, stay relatively consistent with changing temperature. This simplifies control and allows for direct-battery and variable voltage devices. However, starting in 2010, devices began to use onboard processors for software-controlled output power and direct wattage control.

Variable wattage devices can put out consistent vapor even if the resistance of the heating coil changes. Temperature-limiting and temperature-controlled ENDS use heating wires that undergo significant, repeatable changes in resistance when the temperature changes. Typical materials are pure nickel, pure titanium, and stainless steel. The resistance of the wire is an intrinsic property that varies only with temperature, so there is no time lag as happens with an external sensor. A resistance-based temperature-controlled device can react to changes in temperature as fast as the controller can sample the resistance of the coil, often hundreds or thousands of times per second. The measured resistance of the atomizer coil can be used to determine coil temperature, as shown in Fig. 2.5, using the mathematical formula $R_{measured} = R_{ref}[1 + \alpha(T_{actual} - T_{ref})]$, where R_{ref} and T_{ref} are the resistance of the conductor material at a reference temperature, usually 20°C, and α is the temperature coefficient of resistance for the conductor. As of 2016, it is estimated that half of all new large ENDS have some form of temperature control or temperature limiting. However, because of the increased technical sophistication that current implementations require of the user and the relative scarcity of temperature sensing consumables and atomizers, fewer than 20% of devices that are

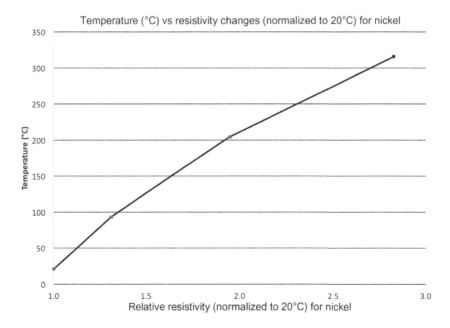

Figure 2.5 Resistance changes in the atomizer coil as a function of temperature.

temperature capable are routinely used in a temperature-controlled mode. This percentage is expected to rise as devices become more user friendly. Further information on this technology and its regulatory perspective is provided in Chapter 5.

The increasing popularity of ENDS is causing concern among public health groups and regulators, given the lack of oversight on these products. In part, this is due to the fact that ENDS manufacturers generally do not provide analytical testing information on device performance or other information such as yield, production of thermal decomposition products, and batch to batch reproducibility. Further, the media and scientific publications tend to regard ENDS as a general class, while a wide range of products are available on the retail market, often with liquid, batteries, and tank selected by consumers from different suppliers.

The scientific literature has shown large performance differences across the range of available ENDS. The impact on nicotine yield and public health is not known.[10] The following sections of this chapter review the available information on the composition of liquids, common contaminants, aerosol production, and compounds of concern that might be present in the aerosol, with emphasis on methods to detect and quantitate.

NICOTINE IN LIQUIDS

Almost all ENDS deliver nicotine. Unlike in tobacco products that include the whole leaf, the amount of added nicotine via ENDS is controlled by the producer of the liquid. Nicotine content is communicated to the consumer in several ways. Prefilled devices may list total nicotine amount or the nicotine concentration of the liquid in units of mg/mL, mg/gram, or percent nicotine by weight. Refill liquids typically list mg/mL or percent nicotine by volume. The yield of aerosol among ENDS varies from less than 2 mg/puff to over 40 mg/puff, depending on the type of device and power level. Nicotine in commercial liquids and prefilled devices varies from 3 mg/mL on the low end to 50 mg/mL on the very high end. **Three to 6** mg/mL nicotine-containing liquids are commonly used in highest yielding ENDS and 25−50 mg/mL in lower yielding prefilled cigalikes. By matching nicotine liquid content to device yield, consumers are exposed to a narrower range of nicotine amount per puff than might be obvious from the range of nicotine content available on the market. Use of a 50 mg/mL liquid would be very unpleasant in a high-yielding fourth-generation ENDS but acceptable in a cigalike. However, the level listed on the labels of ciga-like cartridges and refill solutions is often significantly different from measured values.[11–20] The Salt Lake County Health Department[21] recently tested the amount of nicotine in 153 liquid samples from local retail shops and reported that 61% differed by at least 10% from the labeled content, with discrepancies that ranged from 88% less to 840% more than stated. Of the 33 samples that listed the nicotine amount as zero, 1 contained 7.35 mg/mL. Given the number of reports that have found that nicotine content in the refill liquids or prefilled devices does not match the labeled amounts, it appears that a large percentage of commercial products are not accurately labeled. Table 2.2 summarizes historical nicotine test results for refill solutions; there appears to be little improvement over time in the accuracy of labeling.

However, several industry groups do have standards in place that mandate the nicotine content in refill liquids. The American E-Liquid Manufacturing Standards Association (AEMSA)[24] and the Electronic Cigarette Trade Association of Canada (ECTA)[25] require that their members produce and test liquids within a tolerance level of ± 10% of their labeled nicotine content. Production of refill liquids falling under European Union[26] or US Food and Drug (FDA)[27] regulation will also likely require mandatory nicotine analyses.

There are currently no standardized analysis methods for nicotine in liquids. Several have been used: nonaqueous titrations,[24,28] high-performance liquid chromatography (HPLC),[14] liquid

Table 2.2 Nicotine Reported in Refill Solutions[a]				
Literature	Year	Matrix	Units	Labeled Amount
Goniewicz et al.[11]	2012	Refill solution	mg/mL	0–24
Etter et al.[14]	2013	Refill solution	mg/mL	6–30
Cameron et al.[15]	2012	Refill solution	mg/mL	6–36
Trehy et al.[16]	2011	Refill solution	mg/mL	0–24
Salt Lake[21]	2015	Refill solution	mg/mL	Not provided
Enthalpy[22]	2013	Refill solution	mg/mL	6–36
Peace et al.[23]	2016	Refill solution	mg/mL	6–22
[a]Deviation from label = (measured value − labeled value) × 100/labeled value.				

chromatography—mass spectrometry/mass spectrometry (LC-MS/MS),[15] gas chromatography-flame ionization detector (GC-FID),[11] GC-MS,[14] and GC-nitrogen phosphorous detector (GC-NPD).[29] Reference methods are also available in US Pharmacopeia (USP)[30] and European Pharmacopoeia (Ph. Eur.)[31] monographs, but these do not directly apply to the flavored liquid matrix. Given the range of flavor compounds in liquids, methods must be fully evaluated before systematic adoption. Analysis of nicotine in flavored liquids can be problematic due to the presence of a large number of flavor compounds in samples. These compounds may co-elute and interfere with either nicotine or internal standards, which will cause inaccurate results. Analytical methods based on the use of mass selective detection like GC-MS and LC-MS/MS greatly reduce the possibility of compound misidentification, but these instruments are costlier to purchase and operate than GC-FID or GC-NPD. In our laboratory, GC-FID is the primary method, with the procedure of Smith and Meruva.[32] It has the advantage of being able to also determine PG, GLY, and menthol, and also water if the GC system is equipped with a thermal conductivity detector (TCD). However, since nicotine is usually present at or near the 1% level in liquids, researchers have the option to use HPLC-UV, LC-MS/MS, and GC-MS as well as GC-FID.

COMPOUNDS OF REGULATORY CONCERN LIQUIDS

Combustible tobacco products have a long history of chemical testing linked to product regulation.[33] Testing and reporting of constituents in tobacco and emissions from tobacco products formed the basis of regulatory efforts. Reporting of constituent yields to regulatory bodies is now required in many parts of the world.[26,34] Recent regulations issued by the European Union[26] and proposed regulations by the FDA[27] ensure that ENDS products will also require testing and regulatory reporting for liquid base constituents, such as nicotine, and harmful and potentially harmful constituents (HPHCs) present in liquids and the resulting aerosol emissions. A growing body of reports has been published in the scientific literature documenting the presence of HPHCs in refill solutions and aerosols. The compounds can be grouped into four classes: (1) primary constituents such as PG, GLY, and nicotine; (2) unintended contaminants present in the liquid derived from the primary constituents, flavors, or nicotine used in the production of the liquid, such as tobacco-specific nitrosamines (TSNAs) from the nicotine and ethylene glycol (EG) and diethylene glycol (DEG) from the PG or GLY; (3) thermal decomposition products formed during the heating of the liquid and subsequent formation of the aerosol, including aldehydes and volatile organic compounds (VOCs), and (4) leachable materials that transfer from the device or tank into the liquid. These contaminants may include a range of materials such as plasticizers and metals.[35] While there have been few systematic studies on the transfer of compounds from the liquid into the aerosol and even fewer that determine a wide range of HPHCs in both the liquid and the aerosol, it stands to reason that compounds with physical properties similar to nicotine, PG, or GLY will transfer from the liquid into the aerosol. Further, several compounds that are unlikely to form as the result of thermal decomposition have been found in some aerosol samples—e.g., polyaromatic compounds (PAHs) in tobacco-flavored liquids.[36,37] PAHs are products of combustion[38] and are only formed at temperatures unlikely to be obtained in ENDS;[39] if they are found in ENDS aerosols they should be treated as contaminants that transferred from the ingredients present in the liquid, not as compounds that formed during the production of the aerosol.

Given the different origins of compounds in the liquid and the aerosol, the following two sections will address separately native compounds present in liquid that transfer into the aerosol and compounds that are formed during the generation of the aerosol.

CONTAMINANTS IN LIQUIDS

When the liquid is heated, it and compounds contained in it will be transferred into the aerosol. Compounds with boiling points similar to PG and GLY are thought to transfer in a nearly quantitative manner,[40] while compounds with significantly higher boiling points will likely transfer at lower rates. Little is known about the transfer efficiency of many compounds, but unpublished data from our laboratory show that polar compounds like caffeine do not transfer efficiently. Table 2.3 is a summary of reported compounds found in liquids. The list is broken into four classes based on the suspected origin of each compound.

Contaminants present in the base liquid constituents PG and GLY include EG and DEG, which are known to be toxic to humans and

Table 2.3 Compounds of Concern in ENDS Liquids			
Constituents	TSNAs	Nicotine Related	Others
Nicotine[a]	NNK[a]	Nicotine n-oxide	pH[a]
Propylene glycol[a]	NNN[a]	Cotinine	Diacetyl[a]
Glycerin[a]	NAB	Nornicotine	Acetyl propionyl[a]
Menthol[a]	NAT	Anatabine	Cinnamaldehyde
Water		Myosmine	Butyric acid
Ethylene glycol[a]		Anabasine	Coumarin
Diethylene glycol[a]		Beta-nicotyrine	Phenolics
		Anabasine[a]	Organic acids
			Nitrate
			PAHs[a]
[a]Compounds on the FDA premarket tobacco product application (PMTA) draft list.			

may be present in industrial grades of PG and GLY. In 1999, the FDA reported that DEG was detected in one cartridge at approximately 1%;[12] recently Hutzler et al. found that 5 of 28 samples obtained from the German market contained EG as the major carrier material in the liquid.[12,41] Shan et al. analyzed for EG and DEG in 31 commercial liquids and found EG in 12 and DEG in 8; concentrations ranged from 10 to 143 μg/g for EG and 10 to 20 μg/g for DEG.[42] In our laboratory, we routinely test for EG and DEG in commercial liquids. Since 2014, we have found that approximately 1% of liquids contain DEG at levels above 0.1% but not above 0.3%, while EG has not been detected above 0.1% in any commercial samples. Given the small number of publications on presence of EG and DEG in liquids since 1999, it is likely that the few published reports of high levels of EG and DEG are the results of isolated use of inferior grades of PG and GLY or the mislabeling or mishandling of materials.

Most nicotine used in ENDS is derived from tobacco plants. Only recently did synthetic nicotine become commercially available.[43] While tobacco-based nicotine is highly purified by extraction and distillation, it still contains trace levels of impurities and nicotine-related compounds, avoidance of which is not commercially feasible because of their chemical structures and properties. Known impurities include nicotine-related alkaloids and TSNAs, which are present in all tobacco-derived, nicotine-containing products, including pharmaceutical grade nicotine.[44,45] In 2009, the FDA reported TSNAs in several commercial samples.[12] Goniewicz et al. found N-nitrosonornicotine (NNN) and 4-(methylnitrosamino)1-(3-pyridyl)-1-butanone (NNK) in the aerosol from several ENDS; the content of NNN ranged from 0.8 to 4.3 ng and of NNK from 1.1 to 28.3 ng in 150 puffs.[46] Farsalinos et al. reported that liquids tested contained up to 7.7 ng/g of NNN and up to 2.3 ng/g of N-nitrosoanabasine (NAB), while N'-nitrosoanatabine (NAT) and NNK were not found.[40] TSNAs present in liquids may be imparted due to the use of other tobacco-derived ingredients or materials used in production. Flavors extracted from tobacco could potentially contain additional TSNAs.

The impact of non nicotine sources, e.g., tobacco-extracted flavors, on TSNA content in liquids has not been widely studied. Farsalinos examined refill liquid samples containing tobacco-extracted flavors. The 11 samples had an average of 9.5 and 5.4 ng/mL of NNK and NNN, respectively. In this study, liquids produced from tobacco-derived flavors were compared to liquids produced using conventional flavors, and no evidence was found of an increase in TSNAs in extracted tobacco liquids.[18] TSNAs are very potent carcinogens,[47] but the amounts reported in ENDS are several orders of magnitude lower than those found in combustible products.[48]

In 2009, the FDA reported that tobacco alkaloids anabasine, myosmine, and β-nicotyrine were present in several commercial samples.[12] These compounds occur at relatively high concentrations in tobacco and at low concentrations in purified USP grade nicotine. The USP monograph for nicotine allows for up to 1.0% total impurities, most of which are nicotine-related alkaloids. So tobacco alkaloids in ENDS liquids are not unexpected. These include nornicotine, myosmine, anatabine, and anabasine. Etter et al. reported nicotine-related products present from below the method detection limit to 4.4% of nicotine content; in most of the samples the level was 1−2%, which is outside the limits for pharmaceutical grade nicotine.[14] In addition to myosmine, anatabine and anabasine, *cis*-nicotine oxide and *trans*-nicotine oxide were also reported. Unlike the tobacco-related alkaloids, nicotine oxides are known oxidation products of nicotine and are thought to form during storage. The storage conditions and age of the samples play an important part in the amount of nicotine-related compounds found in liquid and subsequent ENDS aerosol. In 2014, Flora et al. presented the results of long-term storage on nicotine degradants in cigalikes.[49]

Nicotine concentration decreased over time with a corresponding increase in nornicotine, cotinine, and nicotine N-oxides. Also, nornicotine increased from 0.05% to ~0.3% and N-oxides increased from 0.05% to ~0.15% compared to nicotine in the liquids. Since nicotine-related compounds can be present in pure nicotine and can also be formed by degradation of the nicotine over the storage lifetime of ENDS liquids, their measurement in ENDS liquids is necessary. Flora described a straightforward LC-MS/MS analysis method; we use a similar method in our laboratory.

Several other classes of compounds are of concern in liquids. These include a broad category of flavors, organic acids, phenolics, and nitrate. Flavor compounds warrant special attention because of the paucity of inhalation safety data. Studies have shown that the cytotoxic properties of ENDS liquids and aerosol, although significantly lower than those of tobacco smoke, may be attributed to specific flavors,[50–52] an area that certainly requires further research. Several flavor producers and liquid manufacturers attempted to promote flavors as generally recognized as safe (GRAS), but the Flavor and Extract Manufacturers Association of the United States (FEMA) responded negatively:[53]

> *The manufacturers and marketers of e-Cigarettes and all other flavored tobacco products, and flavor manufacturers and marketers, should not represent or suggest that the flavor ingredients used in these products are safe because they have FEMA GRAS™ status for use in food because such statements are false and misleading.*

In addition to the misrepresentation of GRAS status, compounds with known negative health effects have been widely used in some liquids. Diacetyl (DA) and acetyl propionyl (AP) have been particularly common in certain sweet flavor types.[54] These alpha beta diketones have been implicated in the development of bronchiolitis obliterans, an irreversible respiratory disease also called "popcorn lung disease" because it was initially observed in workers exposed to DA at popcorn factories.[55,56] They are subject to pending regulations in the United States and European Union.[26,27] Cinnamaldehyde is used in some cinnamon-flavored liquids and is known to be cytotoxic.[57] Butyric acid is used in some liquids as a replacement for DA and AP, and may have deleterious health effects.[58] Organic acids are also used to reduce the harshness of nicotine.[59]

Tobacco-extracted flavors are produced from whole tobacco leaves. This process may transfer unwanted compounds along with the flavor compounds. The extracts are known to contain elevated levels of phenolic compounds and nitrate[18] and could potentially include PAHs and other combustion products if the source tobacco has been cured with wood smoke, as is dark, fire-cured tobacco.

ANALYTICAL GENERATION OF ENDS AEROSOL

To accurately measure the compounds present in the ENDS aerosol, the aerosol must be generated in a consistent and reproducible manner. This can be done in several ways. It can be generated manually with a syringe and a timer.[60] It can be collected with a semiautomated system consisting of vacuum pumps and flow controllers.[61] It can be collected by analytical puffing machines similar to those used for cigarette smoke.[62] Puffing machines have several advantages, including the ability to accurately control and record puff duration, puff volume, and interpuff interval, and to select from preset puff profiles or custom user-designed puff profiles. Automatic devices are triggered by air flow at the start of each puff, while manual devices must be triggered at the start of each puff. Manual devices can be triggered by hand or interfaced with the puffing machine to automate the process. A wide range of puff profiles have been used to generate ENDS aerosol (see Table 2.6). Two profiles have gained recent acceptance and are now common: a 55-mL puff with a 3-second puff duration and 30-second interpuff delay (55/3/30), which is recommended by the Cooperation Centre for Scientific Research Relative to Tobacco (CORESTA) e-Cigarette Task Force,[63] and 55-mL puff with a 4-second puff duration and 30-second interpuff delay (55/4/30). The latter was reported by Farsalinos to represent the average puff parameters in a study of experienced ENDS users.[64]

Once the aerosol is generated, the sample must be trapped for analysis, and here too several approaches are common. A standard Cambridge filter designed for tobacco smoke has been shown to collect >98% of nicotine, PG, and GLY from ENDS aerosol.[65] Compounds more volatile than those three are trapped by specialized systems such as liquid-filled impingers[62] and sorbent media.[66]

IMPACT OF PUFF PROFILE ON ENDS YIELD

One of the largest differences between ENDS and combustible cigarettes is the effect of puff volume and puff duration on overall yield of smoke and aerosol constituents. Combustible cigarettes, by their very design, require a supply of oxygen to burn efficiently. The more oxygen supplied to the coal, the more tar and other combustion products are delivered to the smoker. The yield of a cigarette can be predicted if the total volume of air that passes through the burning coal is known.[67] Puff duration and puff profile have limited effect on the overall yield of combustible products. In contrast to combustible cigarettes, ENDS puff volume and puff flow rate have little to no effect on overall yield.[68,69] Instead, puff duration is the major factor. Since ENDS contain a heated atomizer coil, increasing the length of time that the wire is heated, increases the amount of liquid that is aerosolized during each puff. The effect of puff duration is linear, with a non-zero intercept, because the coil must reach operating temperature before aerosol can be produced. After the aerosolization temperature is reached, and provided that sufficient liquid is supplied to the coil during heating, aerosol yield will increase with increased puff duration. Puff flow rate does not have a major impact on yield except with automatic, air flow-activated devices since these devices require a minimum airflow for activation.

Cigalike ENDS contain a pressure sensor to detect the start of puffing. Flow rates below the activation point of the sensor will not produce aerosol. The use of a constant flow or "square wave" puff profile is recommended when testing devices that have pressure sensors. The use of non constant flow puff profiles such as bell or triangular shaped is not recommended with automatic devices, because the flow rate is greatly reduced at the beginning of the puff and changes during the puff. Those profiles can produce inconsistent activation, resulting in shorter than expected time when the coil is heated. Fig. 2.6 shows results from identical devices with different profiles. In panel A, 55-mL puff volume over 4 seconds with constant flow rate was compared to 35-mL puff volume over 2 seconds with bell-shaped flow rate for effect on per-puff yield. In panel B, 55-mL puff volume over 4 seconds with constant flow rate, 70-mL puff volume over 3 seconds with constant flow rate, and 55-mL puff volume over 2 seconds with bell-shaped flow rate were compared for effect on cumulative yield.

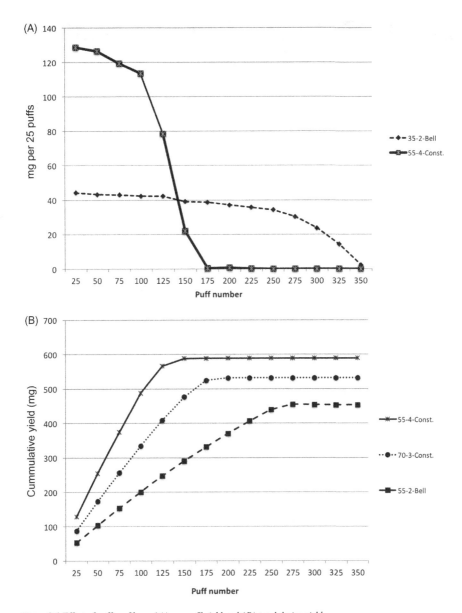

Figure 2.6 Effect of puff profile on (A) per-puff yield and (B) total device yield.

COMPOUNDS OF CONCERN IN ENDS AEROSOL

The ENDS user inhales an aerosol comprised of the base liquid ingredients (typically PG and GLY), nicotine, and flavors. Leachable materials present in the device may transfer into the liquid and the resulting

Table 2.4 Compounds of Concern in ENDS Aerosol			
Constituents and Contaminants	Thermal Decomposition	Combustion Products	Leachables
Nicotine[a]	Acetaldehyde[a]	1-Aminonaphthalene[a]	Arsenic
Propylene glycol[a]	Acrolein[a]	2-Aminonaphthalene[a]	Cadmium
Glycerin[a]	Crotonaldehyde[a]	4-Aminobiphenyl[a]	Chromium
Menthol[a]	Formaldehyde[a]	Benzo[a]pyrene[a]	Lead[a]
Water		1,3-Butadiene[a]	Nickel[a]
Ethylene glycol[a]		Acrylonitrile[a]	Zinc
Diethylene glycol[a]		Benzene[a]	
TSNAs[a]		Isoprene[a]	
Anabasine[a]		Toluene[a]	
[a]Compounds on the FDA premarket tobacco product application (PMTA) list.			

aerosol, including metals from the tank and heating coil. It is also possible that new compounds are formed during the heating and aerosolization of the liquid. In this case, there will be compounds present in the aerosol that are either not present in the starting liquid or increase in concentration during heating. In very general terms, these compounds fall into four classes, as shown in Table 2.4.

The first class are the primary constituents or minor impurities present in the liquid before heating. These include nicotine, PG, GLY, water, EG, DEG, alkaloids, and TSNAs. Nearly all scientific publications to date have shown that these liquid constituents will transfer from the liquid to the aerosol,[40,70−72] but that the amount transferred will depend on many factors and must be determined for each device tested.[62,73] Minor impurities such as EG, DEG, and anabasine are not likely to be produced during the heating of the liquid to form the aerosol. Testing of these compounds in the liquid, not the aerosol, may be more appropriate, given their origin. Analytical methods for liquid samples are also suitable for aerosol samples, provided that trapping efficiency and recovery from the trapping system is also verified.

The second class are the thermal decomposition compounds produced during the heating of PG and GLY. Aldehydes, including formaldehyde, acetaldehyde, and acrolein, are known to form[74−78] and are of concern; formaldehyde is classified by the International Agency for Research on Cancer (IARC) as a human carcinogen (Group 1) and acetaldehyde as possibly carcinogenic to humans (Group 2B).[79] Acrolein causes irritation of the nasal cavity and damages the lining of

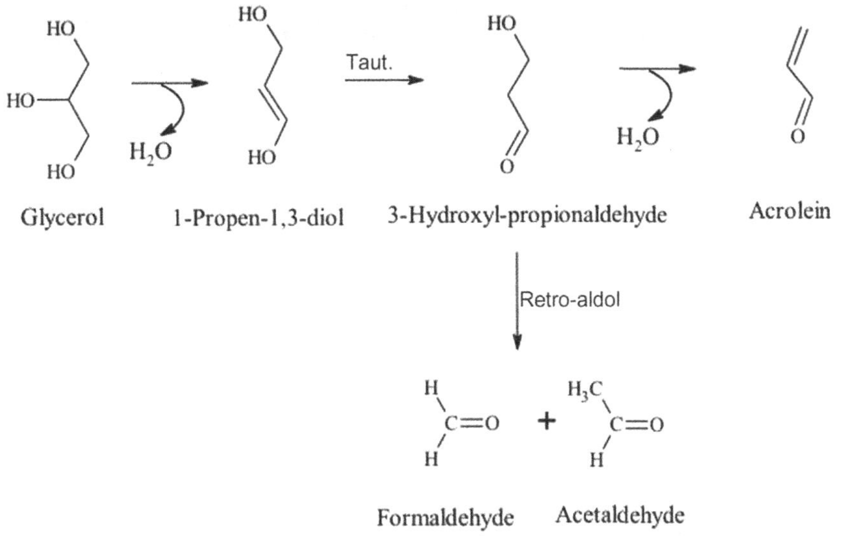

Figure 2.7 The pyrolytic reactions of GLY to produce formaldehyde, acetaldehyde, and acrolein. Radical intermediates for steps involving loss of water are omitted for simplicity.

the lung.[80] Glycerin has been shown to produce these three aldehydes by thermal decomposition (pyrolysis) in temperature-dependent amounts,[81] with small amounts formed in some ionic environments at 350°C and all three formed at 600°C. Fig. 2.7 shows the pathway for this pyrolysis. It involves a free-radical dehydration of GLY to form 3-hydroxyl-1-propen-1-ol, which tautomerizes to 3-hydroxylpropionaldehyde. This then loses another water in a free-radical mechanism to form acrolein. At higher temperatures, 3-hydroxylpropionaldehyde can convert to formaldehyde and acetaldehyde by way of a retro-aldol reaction, which easily cleaves the C2−C3 bond at >400°C.

Table 2.5 is a summary of published studies on aldehydes in ENDS aerosol. On a per-puff basis, the values span an incredible range, from less than 0.01 μg/puff to almost 100 μg/puff for formaldehyde alone. The wide range does not appear to be related to differences in analytic methods, since trapping with DNPH followed by HPLC separation is commonly used. Instead, they depend on product design and power applied to the coil. Applying too much power to the atomizer will result in overheating of the coil because of insufficient liquid supply. Under these dry puff conditions, coil temperatures may exceed 300°C[82] and can lead to the production of thermal decomposition or pyrolysis products. Devices tested under these

Table 2.5 Summary of Published Studies on Aldehydes in ENDS Aerosol			
Authors	Methods of Generation	Types of ENDS	Carbonyl Emissions Per Puff (μg)
Sleiman et al.[61]	Vacuum pump (50 mL, 5.0 s, 25 s, 5 puffs), DNPH-sorbent, HPLC	Third-generation e-Cigarette with variable voltage (3.3–4.8 V)	Formaldehyde 53–97
			Acetaldehyde 10–50
			Acrolein 3–21.5
Flora et al.[72]	Machine smoking (55 mL, 4.0 s, 30 s, 20 puffs), impinger with DNPH, HPLC, UPLC	Four commercial cigalike products	Formaldehyde 0.090–0.33
			Acetaldehyde <0.71
			Acrolein <0.09
Gillman et al.[62]	Machine smoking (55 mL, 4.0 s, 30 s, 25 puffs), impinger with DNPH, HPLC	Third-generation e-Cigarette with variable voltage (3.8–5.0 V) or wattage (10–25 W)	Formaldehyde 0.05–51
			Acetaldehyde 0.03–40.7
			Acrolein <0.02–5.5
Geiss et al.[82]	Machine smoking (50 mL, 3.0 s, 30 s, 10 puffs), DNPH-sorbent, HPLC	Third-generation e-Cigarette with variable wattage (5–25 W)	Formaldehyde 0.024–1.6
			Acetaldehyde 0.013–0.35
			Acrolein ND to 0.025
Tayyarah and Long[36]	Machine smoking (55 mL, 2.0 s, 30 s, 99 puffs), aerosol collected in two impingers with DNPH, HPLC	Two disposable and three rechargeable commercial cigalike products	Formaldehyde <0.01
			Acetaldehyde <0.06–0.32
			Acrolein <0.02–0.19
Kosmider et al.[83]	Machine smoking (70 mL, 1.8 s, 17 s, 30 puffs), DNPH-sorbent tubes, HPLC	Second-generation e-Cigarette with variable voltage (3.2–4.8 V)	Formaldehyde[a] 0.01–1.8
			Acetaldehyde[a] 0.01–0.28
			[a]Average values
Uchiyama et al.[75]	Machine smoking (55 mL, 2 s, 30 s, 10 puffs), DNPH-sorbent tubes, HPLC	Thirteen Japanese commercial e-Cigarette products	Formaldehyde[a] ND to 3.4
			Acetaldehyde[a] ND to 2.6
Bekki et al.[84]			Acrolein[a] ND to 2.0
			[a]Average values

(Continued)

Table 2.5 (Continued)			
Authors	Methods of Generation	Types of ENDS	Carbonyl Emissions Per Puff (μg)
Goniewicz et al.[46]	Machine smoking (70 mL, 1.8 s, 10 s, 15 puffs), sorbent trapping, HPLC	11 popular Polish commercial cigalike products	Formaldehyde 0.01−0.37
			Acetaldehyde 0.001−0.09
			Acrolein 0.005−0.28
Uchiyama et al.[85]	Machine smoking (55 mL, 2 s, 30 s, 10 puffs), DNPH-sorbent tubes, HPLC	Second-generation e-Cigarette	Formaldehyde[a] 0.19−1
			Acetaldehyde[a] 0.14−7.3
			Acrolein[a] ND to 1.5
			[a]Average values

[a]*Puffing conditions are given in the following format: puff volume, puff duration, time between puffs, and number of puffs per sample.*

conditions can produce unpalatable, acrid aerosol and show artificially high aldehyde values.[86]

The third class of compounds is more difficult to classify, because the mechanisms by which they are produced are currently not known and they are more commonly found in combustible products. These compounds include (1) polyaromatic amines (PAAs),[36] specifically aminonaphthalenes and aminobiphenyl, (2) VOCs[46,60,61] such as benzene and isoprene, and (3) PAHs,[37] including benzo[a]pyrene. PAAs and PAHs are on the FDA Draft PMTA list,[27] but their presence in ENDS aerosols is not supported by recent literature. Few methods are published for the analysis of PAAs, PAHs, and VOCs in ENDS aerosol. Flora et al.[72] and Tayyarah and Long[36] applied traditional tobacco smoke analysis with good success. Two recent studies[60,61] applied thermal desorption coupled to GC-MS for the determination of VOCs. This approach offers a very low detection limit and also the ability to identify new compounds based on comparison to reference libraries.

The fourth class is contaminants introduced from the device into the liquid, including metals, before or during the formation of the aerosol. These should be considered leachables. Goniewicz et al.[46]

reported cadmium, chromium, nickel, and arsenic in aerosol. Williams et al.[35] reported particles larger than 1 micron comprised of tin, silver, iron, nickel, aluminum, and silicate, and nanoparticles of less than 0.10 micron of tin, chromium, and nickel in aerosol. Concentrations of 9 of 11 elements were higher than or equal to the corresponding concentrations in cigarette smoke. The presence of metals in ENDS aerosol is not surprising, given that the heating coil and often the tank material are constructed of metal. Most published studies used inductively coupled plasma mass spectrometry (ICP-MS), which provides the selectivity and sensitivity required. Beauval et al.[87] recently published an ICP-MS method for the analysis of metals in refill liquids, and Goniewicz et al.[46] have published methods for the collection and analysis of metals in aerosol. In our laboratory, we collect aerosol samples using either electrostatic precipitation in quartz tubes or on metal-free filter pads such as quartz, followed by ICP-MS.

CONCLUSION

The science of ENDS is evolving rapidly, especially compared to the science of combustible tobacco. The range of products and their customizability complicates testing by presenting researchers with a practically unlimited range of choices. The lack of research data generated under actual product usage conditions and the paucity of information from device producers means that test results do not adequately reflect human usage. However, recent work is closing the knowledge gaps, and more researchers are now taking care to test under appropriate conditions instead of conditions that give the highest possible yield of potentially harmful compounds. Validated, standardized methods are still needed for the most basic of ENDS testing. In the future, standardized methods for the machine generation of ENDS aerosol and methods for the analysis of emissions will undoubtedly be available to all researchers. Until that time, the appropriateness of testing conditions and test methods must be evaluated on a case-by-case basis.

REFERENCES

1. Hon L, Inventor. Electronic atomization cigarette. US patent US2007267031; November 11, 2007.

2. Gilbert HA, Inventor. Smokeless non-tobacco cigarette. US patent 3,200,819; August 17, 1965.

3. Regan AK, Promoff G, Dube SR, Arrazola R. Electronic nicotine delivery systems: adult use and awareness of the "e-cigarette" in the USA. *Tob Control* 2013;**22**(1):19–23.

4. Pearson JL, Richardson A, Niaura RS, Vallone DM, Abrams DB. e-Cigarette awareness, use, and harm perceptions in US adults. *Am J Public Health* 2012;**102**(9):1758–66.

5. Vardavas CI, Filippidis FT, Agaku IT. Determinants and prevalence of e-cigarette use throughout the European Union: a secondary analysis of 26566 youth and adults from 27 countries. *Tob Control* 2014;**0**:1–7.

6. Barbeau AM, Burda J, Siegel M. Perceived efficacy of e-cigarettes versus nicotine replacement therapy among successful e-cigarette users: a qualitative approach. *Addict Sci Clin Pract* 2013;**8**(1):5.

7. Pauly J, Li Q, Barry MB. Tobacco-free electronic cigarettes and cigars deliver nicotine and generate concern. *Tob Control* 2007;**16**(5):357.

8. Trtchounian A, Williams M, Talbot P. Conventional and electronic cigarettes (e-cigarettes) have different smoking characteristics. *Nicotine Tob Res* 2010;**12**(9):905–12.

9. Talih S, Balhas Z, Salman R, Karaoghlanian N, Shihadeh A. "Direct dripping": a high-temperature, high-formaldehyde emission electronic cigarette use method. *Nicotine Tob Res* 2016;**18**(4):453–9.

10. Chen IL. FDA summary of adverse events on electronic cigarettes. *Nicotine Tob Res* 2013;**15**(2):615–16.

11. Goniewicz ML, Kuma T, Gawron M, Knysak J, Kosmider L. Nicotine levels in electronic cigarettes. *Nicotine Tob Res* 2013;**15**(1):158–66.

12. Westenberger BJ. *Evaluation of e-cigarettes*. Washington, DC: U.S. Food and Drug Administration; 2009. Available from: <http://www.fda.gov/downloads/Drugs/ScienceResearch/UCM173250.pdf> [accessed June 2016].

13. Williams M, Talbot P. Variability among electronic cigarettes in the pressure drop, airflow rate, and aerosol production. *Nicotine Tob Res* 2011;**13**(12):1276–83.

14. Etter JF, Zather E, Svensson S. Analysis of refill liquids for electronic cigarettes. *Addiction* 2013;**108**(9):1671–9.

15. Cameron JM, Howell DN, White JR, Andrenyak DM, Layton ME, Roll JM. Variable and potentially fatal amounts of nicotine in e-cigarette nicotine solutions. *Tob Control* 2014;**23**(1):77–8.

16. Trehy ML, Ye W, Hadwiger ME, et al. Analysis of electronic cigarette cartridges, refill solutions, and smoke for nicotine and nicotine related impurities. *J Liq Chromatogr Related Technol* 2011;**34**(14):1442–58.

17. Cobb NK, Byron MJ, Abrams DB, Shields PG. Novel nicotine delivery systems and public health: the rise of the "e-cigarette". *Am J Public Health* 2010;**100**(12):2340–2.

18. Farsalinos KE, Gillman IG, Melvin MS, et al. Nicotine levels and presence of selected tobacco-derived toxins in tobacco flavoured electronic cigarette refill liquids. *Int J Environ Res Public Health* 2015;**12**(4):3439–52.

19. Lisko JG, Tran H, Stanfill SB, Blount BC, Watson CH. Chemical composition and evaluation of nicotine, tobacco alkaloids, pH, and selected flavors in e-cigarette cartridges and refill solutions. *Nicotine Tob Res* 2015;**17**(10):1270–8.

20. Cheah NP, Chong NW, Tan J, Morsed FA, Yee SK. Electronic nicotine delivery systems: regulatory and safety challenges: Singapore perspective. *Tob Control* 2014;**23**(2):119–25.

21. *Nicotine content in e-liquid samples*. Salt Lake County Health Department. https://ibis.health.utah.gov/pdf/opha/publication/hsu/2015/1501_HPV.pdf [accessed June 2016].

22. Melvin MS, Gillman G, Humphries KE. Aerosol production and chemical analysis of electronic cigarettes using a linear smoking machine. Paper presented at: *67th Tobacco Science*

Research Conference, September 18–21, 2013, Williamsburg, VA. Available from: https://www.coresta.org/sites/default/files/events/67th-TSRC-Abstracts_2013.pdf [accessed June 2016].

23. Peace MR, Baird TR, Smith N, Wolf CE, Poklis JL, Poklis A. Concentration of nicotine and glycols in 27 electronic cigarette formulations. *J Anal Toxicol* 2016;**40**(6):403–7.

24. *E-Liquid Manufacturing Standards, 2015, v2.3*. American E-liquid Manufacturing Standards Association. http://www.aemsa.org/wp-content/uploads/2015/09/AEMSA-Standards-v2.3.pdf [accessed June 2016].

25. *E-Liquid Testing Standards*. Electronic Cigarette Trade Association of Canada. http://ectaofcanada.com/pagedisp.php?section=E-Liquid_Testing [accessed June 2016].

26. European Union. On the approximation of the laws, regulations and administrative provisions of the Member States concerning the manufacture, presentation and sale of tobacco and related products and repealing. *Official Journal of the European Union* 2014.

27. Guidance for Industry. *Premarket tobacco product applications for electronic nicotine delivery systems*. U.S. Food and Drug Administration. Available from: http://www.fda.gov/downloads/TobaccoProducts/Labeling/RulesRegulationsGuidance/UCM499352.pdf; 2016 [accessed June 2016].

28. Vieira CA, de Paiva SA, Funai MN, Bergamaschi MM, Queiroz RH, Giglio JR. Quantification of nicotine in commercial brand cigarettes: how much is inhaled by the smoker? *Biochem Mol Biol Educ* 2010;**38**(5):330–4.

29. McAuley TR, Hopke PK, Zhao J, Babaian S. Comparison of the effects of e-cigarette vapor and cigarette smoke on indoor air quality. *Inhal Toxicol* 2012;**24**(12):850–7.

30. United States Pharmacopoeial and National Formulary. USP 35-NF 30. Rockville, MD: United States Pharmacopoeial Convention; 2012.

31. European Pharmacopeia. *Nicotine monograph 1452*. Strasbourg, France: Council of Europe; 2001.

32. Smith T, Meruva N. Simultaneous determination of six e-cigarette formulation components by GC. Paper presented at: CORESTA Congress, October 12–16, 2014, Quebec, Canada. Available from: http://www.altria.com/ALCS-Science/ConferenceDocumentLibrary/Smith-Simultaneous determination of six e-cigarette formulation components by GC.pdf [accessed June 2016].

33. Institute of Medicine Committee to assess the science base for tobacco harm reduction. In: Stratton K, Shetty P, Wallace R, Bondurant S, editors. *Clearing the smoke: assessing the science base for tobacco harm reduction*. Washington, DC: National Academies Press (US); 2001.

34. Guidance for industry. *Reporting harmful and potentially harmful constituents in tobacco products as used in Section 904(e) of the Federal Food, Drug, and Cosmetics Act*. U.S. Food and Drug Administration 2011. Available from: http://www.fda.gov/downloads/TobaccoProducts/GuidanceComplianceRegulatoryInformation/UCM241352.pdf [accessed June 2016].

35. Williams M, Villarreal A, Bozhilov K, Lin S, Talbot P. Metal and silicate particles including nanoparticles are present in electronic cigarette cartomizer fluid and aerosol. *PLoS One* 2013;**8**(3):e57987.

36. Tayyarah R, Long GA. Comparison of select analytes in aerosol from e-cigarettes with smoke from conventional cigarettes and with ambient air. *Reg Tox Pharm* 2014;**70**(3):704–10.

37. Laugesen M. Safety report on the Ruyan® e-cigarette cartridge and inhaled aerosol; 2008. http://www.healthnz.co.nz/RuyanCartridgeReport30-Oct-08.pdf [accessed June 2016].

38. Kislov VV, Sadovnikov AI, Mebel AM. Formation mechanism of polycyclic aromatic hydrocarbons beyond the second aromatic ring. *J Phys Chem A* 2013;**117**(23):4794–816.

39. Zhao T, Shu S, Guo Q, Zhu Y. Effects of design parameters and puff topography on heating coil temperature and mainstream aerosols in electronic cigarettes. *Atmos Environ* 2016;**134**:61–9.

40. Farsalinos KE, Gillman G, Poulas K, Voudris V. Tobacco-specific nitrosamines in electronic cigarettes: comparison between liquid and aerosol levels. *Int J Environ Res Public Health* 2015;**12**(8):9046−53.

41. Hutzler C, Paschke M, Kruschinski S, Henkler F, Hahn J, Luch A. Chemical hazards present in liquids and vapors of electronic cigarettes. *Arch Toxicol* 2014;**88**(7):1295−308.

42. Shan NH, McFarlane C, Wagner KA, Flora JW. Quantitative screening of potential contaminants in e-cigarette formulations: ethylene glycol and diethylene glycol. Paper presented at: *CORESTA smoke science and product technology*, October 4−8, 2015, Jeju Island, South Korea. Available from: http://www.altria.com/ALCS-Science/ConferenceDocumentLibrary/TSRC72-EG-DEG-20150903.pdf [accessed June 2016].

43. Next Generation Labs. http://www.nextgenerationlabs.com [accessed June 2016].

44. U.S. Pharmcopeial Forum. *Nicotine*. Rockville, MD: United States Pharmacopoeial Convention; 2006.

45. Kim HJ, Shin HS. Determination of tobacco-specific nitrosamines in replacement liquids of electronic cigarettes by liquid chromatography−tandem mass spectrometry. *J Chromatogr A* 2013;**1291**:48−55.

46. Goniewicz ML, Knysak J, Gawron M, et al. Levels of selected carcinogens and toxicants in vapour from electronic cigarettes. *Tob Control* 2014;**23**(2):133−9.

47. Hecht SS, Hoffmann D. Tobacco-specific nitrosamines, an important group of carcinogens in tobacco and tobacco smoke. *Carcinogenesis* 1988;**9**(6):875−84.

48. Counts ME, Morton MJ, Laffoon SW, Cox RH, Lipowicz PJ. Smoke composition and predicting relationships for international commercial cigarettes smoked with three machine-smoking conditions. *Reg Tox Pharm* 2005;**41**(3):185−227.

49. Flora JW, Miller JH, Schrall J, Smith J, McFarlane C, Meruva N. Nicotine and related impurities in e-cigarette cartridges: stability studies and methodologies. Paper presented at: *CORESTA Congress*, October 12−16, 2014, Quebec. Available from: http://www.altria.com/ALCS-Science/ConferenceDocumentLibrary/Flora Nicotine and related impurities CORESTA 2014.pdf [accessed June 2016].

50. Farsalinos K, Romagna G, Allifranchini E, et al. Comparison of the cytotoxic potential of cigarette smoke and electronic cigarette vapour extract on cultured myocardial cells. *Int J Environ Res Publ Health* 2013;**10**(10):5146.

51. Romagna G, Allifranchini E, Bocchietto E, Todeschi S, Esposito M, Farsalinos KE. Cytotoxicity evaluation of electronic cigarette vapor extract on cultured mammalian fibroblasts (ClearStream-LIFE): comparison with tobacco cigarette smoke extract. *Inhal Toxicol* 2013;**25**(6):354−61.

52. Bahl V, Lin S, Xu N, Davis B, Wang Y-h, Talbot P. Comparison of electronic cigarette refill fluid cytotoxicity using embryonic and adult models. *Reprod Toxicol* 2012;**34**(4):529−37.

53. FEMA. Safety assessment and regulatory authority to use flavors: focus on e-cigarettes. Available from: http://www.femaflavor.org/safety-assessment-and-regulatory-authority-use-flavors-focus-e-cigarettes; 2016 [accessed June 2016].

54. Farsalinos KE, Kistler KA, Gillman G, Voudris V. Evaluation of electronic cigarette liquids and aerosol for the presence of selected inhalation toxins. *Nicotine Tob Res* 2015;**17**(2):168−74.

55. Kreiss K, Gomaa A, Kullman G, Fedan K, Simoes EJ, Enright PL. Clinical bronchiolitis obliterans in workers at a microwave-popcorn plant. *N Engl J Med* 2002;**347**(5):330−8.

56. Kanwal R, Kullman G, Piacitelli C, et al. Evaluation of flavorings-related lung disease risk at six microwave popcorn plants. *J Occup Environ Med* 2006;**48**(2):149−57.

57. Behar RZ, Davis B, Wang Y, Bahl V, Lin S, Talbot P. Identification of toxicants in cinnamon-flavored electronic cigarette refill fluids. *Toxicol In Vitro* 2014;**28**(2):198−208.

58. Cueno ME, Kamio N, Seki K, Kurita-Ochiai T, Ochiai K. High butyric acid amounts induce oxidative stress, alter calcium homeostasis, and cause neurite retraction in nerve growth factor-treated PC12 cells. *Cell Stress Chaperones* 2015;**20**(4):709–13.

59. Bowen A, Xing C, Inventors. PAX Labs, Inc., assignee. Nicotine salt formulations for aerosol devices and methods thereof. US Patent 9215895; December 22, 2015.

60. Herrington JS, Myers C. Electronic cigarette solutions and resultant aerosol profiles. *J Chromatogr A* 2015;**1418**:192–9.

61. Sleiman M, Logue JM, Montesinos VN, et al. Emissions from electronic cigarettes: key parameters affecting the release of harmful chemicals. *Environ Sci Technol* 2016.

62. Gillman IG, Kistler KA, Stewart EW, Paolantonio AR. Effect of variable power levels on the yield of total aerosol mass and formation of aldehydes in e-cigarette aerosols. *Reg Tox Pharm* 2016;**75**:58–65.

63. CORESTA. CORESTA Recommended Method No. 81: Routine analytical machine for e-cigarette aerosol generation and collection – definitions and standard. Available from: www.coresta.org/sites/default/files/technical_documents/main/CRM_81.pdf; 2015 [accessed June 2016].

64. Farsalinos K, Romagna G, Tsiapras D, Kyrzopoulos S, Voudris V. Evaluation of electronic cigarette use (vaping) topography and estimation of liquid consumption: implications for research protocol standards definition and for Public Health Authorities' Regulation. *Int J Environ Res Publ Health* 2013;**10**(6):2500–14.

65. Alderman SL, Song C, Moldoveanu SC, Cole SK. Particle size distribution of e-cigarette aerosols and the relationship to cambridge filter pad collection efficiency. *Beitr Tabakforsch* 2015;**26**(4):183–90.

66. Uchiyama S, Inaba Y, Kunugita N. Determination of acrolein and other carbonyls in cigarette smoke using coupled silica cartridges impregnated with hydroquinone and 2,4-dinitrophenylhydrazine. *J Chromatogr A* 2010;**1217**(26):4383–8.

67. Browne CL, Keith CH, Allen RE. The effect of filter ventilation on the yield and composition of mainstream and sidestream smokes. *Beitr Tabakforsch* 1980;**10**(2):81–90.

68. Talih S, Balhas Z, Eissenberg T, et al. Effects of user puff topography, device voltage, and liquid nicotine concentration on electronic cigarette nicotine yield: measurements and model predictions. *Nicotine Tob Res* 2015;**17**(2):150–7.

69. Miller JH, Wilkinson J, Flora JW. Effect of puff duration and puff volume on e-cigarette aerosol collection. Paper presented at: *CORESTA smoke science and product technology*, October 4–8, 2015, Jeju Island, South Korea. Available from: http://www.altria.com/ALCS-Science/ConferenceDocumentLibrary/TSRC69-Puffs-20150915.pdf; 2015 [accessed June 2016].

70. Farsalinos KE, Yannovits N, Sarri T, Voudris V, Poulas K. Protocol proposal for, and evaluation of, consistency in nicotine delivery from the liquid to the aerosol of electronic cigarettes atomizers: regulatory implications. *Addiction* 2016;**111**(6):1069–76.

71. Guthery W. Emissions of toxic carbonyls in an electronic cigarette. *Beitr Tabakforsch* 2016;**27**(1).

72. Flora JW, Meruva N, Huang CB, et al. Characterization of potential impurities and degradation products in electronic cigarette formulations and aerosols. *Reg Tox Pharm* 2016;**74**:1–11.

73. Pagano T, DiFrancesco AG, Smith SB, et al. Determination of nicotine content and delivery in disposable electronic cigarettes available in the United States by gas chromatography–mass spectrometry. *Nicotine Tob Res* 2016;**18**(5):700–7.

74. Paschke T, Scherer G, Heller WD. Effects of ingredients on cigarette smoke composition and biological activity: a literature overview. *Beitr Tabakforsch* 2014;**20**(3):107–247.

75. Uchiyama S, Ohta K, Inaba Y, Kunugita N. Determination of carbonyl compounds generated from the e-cigarette using coupled silica cartridges impregnated with hydroquinone and 2,4-dinitrophenylhydrazine, followed by high-performance liquid chromatography. *Anal Sci* 2013;**29**(12):1219−22.

76. Ohta K, Uchiyama S, Inaba Y, Nakagome H, Kunugita N. Determination of carbonyl compounds generated from the electronic cigarette using coupled silica cartridges impregnated with hydroquinone and 2,4-dinitrophenylhydrazine. *Bunseki Kagaku* 2011;**60**:791−7.

77. Lauterbach JH, Spencer A. Generation of acetaldehyde and other carbonyl compounds during vaporization of glycerol and propylene glycol during puffing of a popular style of e-cigarette. Society of Toxicology National Meeting, San Diego, CA; 2015.

78. Flora JW, Wilkinson CT, Wilkinson J, Miller JH. Sensitive and selective method for carbonyl determination in e-cigarette aerosols. Paper presented at: CORESTA Smoke Science and Product Technology, October 4−8, 2015, Jeju Island, South Korea. Available from: http://www.altria.com/ALCS-Science/ConferenceDocumentLibrary/TSRC52.pdf   [accessed June 2016].

79. IARC. *IARC monographs on the evaluation of carcinogenic risks to humans: agents classified by the IARC Monographs,* Volumes 1−113. Geneva, Switzerland: International Agency for Research on Cancer (IARC); 2012.

80. USEPA. *Toxicological review of acrolein.* Washington, DC: U.S. Environmental Protection Agency (U.S. EPA); 2003.

81. Paine JB, Pithawalla YB, Naworal JD, Thomas Jr CE. Carbohydrate pyrolysis mechanisms from isotopic labeling: Part 1: The pyrolysis of glycerin: discovery of competing fragmentation mechanisms affording acetaldehyde and formaldehyde and the implications for carbohydrate pyrolysis. *J Anal Appl Pyrolysis* 2007;**80**(2):297−311.

82. Geiss O, Bianchi I, Barrero-Moreno J. Correlation of volatile carbonyl yields emitted by e-cigarettes with the temperature of the heating coil and the perceived sensorial quality of the generated vapours. *Int J Hyg Environ Health* 2016;**219**(3):268−77.

83. Kosmider L, Sobczak A, Fik M, et al. Carbonyl compounds in electronic cigarette vapors: effects of nicotine solvent and battery output voltage. *Nicotine Tob Res* 2014;**16**(10): 1319−26.

84. Bekki K, Uchiyama S, Ohta K, Inaba Y, Nakagome H, Kunugita N. Carbonyl compounds generated from electronic cigarettes. *Int J Environ Res Public Health* 2014;**11**(11):11192−200.

85. Uchiyama S, Senoo Y, Hayashida H, Inaba Y, Nakagome H, Kunugita N. Determination of chemical compounds generated from second-generation e-cigarettes using a sorbent cartridge followed by a two-step elution method. *Anal Sci* 2016;**32**(5):549−55.

86. Farsalinos KE, Voudris V, Poulas K. E-cigarettes generate high levels of aldehydes only in "dry puff" conditions. *Addiction* 2015;**110**(8):1352−6.

87. Beauval N, Howsam M, Antherieu S, et al. Trace elements in e-liquids—development and validation of an ICP-MS method for the analysis of electronic cigarette refills. *Reg Tox Pharm* 2016;**79**:144−8.

CHAPTER 3

Exposures to e-Cigarette Vapor

J. Thornburg

This chapter presents a review of published data and new data collected by RTI International to provide the reader a perspective on the sources and external factors that influence the magnitude of exposure to e-Cigarette vapor. The author reviewed over 100 published articles from the e-Cigarette, conventional cigarette, and aerosol science fields. Articles were selected that have data to help advance our understanding of user, secondary, and tertiary exposures. Although the cited literature in this chapter is limited, the number of published papers focused on e-Cigarette vapor exposures continues to increase annually. As such, the literature review was limited to those published before August 1, 2016.

A note on terminology: Vapors includes both the particles and gases produced by an e-Cigarette. Particles and aerosols are used synonymously following the definition by Hinds: "An aerosol is defined in its simplest form as a collection of solid or liquid particles in a gas."[1]

INTRODUCTION

This chapter discusses the physical and chemical changes in e-Cigarette vapor during inhalation and exhalation by the user. The vapor properties span aerosol size distribution, the fraction of a chemical constituent in the aerosol and gas phases, and the chemical concentration. These factors influence the nicotine and flavoring dose received by the user and exhaled leading to secondhand exposure. The physical and chemical processes will be supported by published theoretical and empirical evidence plus new findings from current research. How the user and device profile influences the chemical and physical properties of the produced vapors is discussed in Chapter 2. Although that is briefly mentioned in this chapter to add supplemental information, the reader

Analytical Assessment of e-Cigarettes. DOI: http://dx.doi.org/10.1016/B978-0-12-811241-0.00003-6

should refer to Chapter 2 for details. The toxicological and physiological impacts of e-Cigarette vapors are discussed in Chapter 4.

e-CIGARETTE CONSTITUENT APPORTIONMENT

e-Cigarette liquids (e-liquids) are chemically complex. The main ingredients by mass are a carrier liquid (typically propylene glycol [PG] and/ or vegetable glycerin [VG]), water, and nicotine. Each liquid is then customized with a unique combination of flavorings, preservatives, and artificial colors selected by the manufacturer. How these ingredients apportion from the liquid into the aerosol and gas phases during e-Cigarette use has not been extensively studied. Table 3.1 gives the available gas–particle apportionment data for selected e-liquid ingredients. In general, the apportionment of the constituents between the particle and gas phases is dependent on the vapor pressure of each constituent. A chemical with a higher vapor pressure will be more volatile and more likely to prefer the gas phase once heated by the e-Cigarette device. The polar volatile compounds, however, will partition between the gas and particle phases because the PG/VG carriers are humectants. Their affinity for polar compounds will retain

Table 3.1 Qualitative Distribution of Different e-Cigarette Constituents Across the Particle and Gas Phases

Constituents	Liquid	Particle Phase[a]	Gas Phase	Purpose	Source
Propylene glycol	Y	Y	Y	Carrier	Thornburg et al.[2]; Schober et al.[3]
Glycerin	Y	Y	Y	Carrier	Thornburg et al.[2]; Schober et al.[3]
Nicotine	Y	Y	Y	Stimulant	Thornburg et al.[2]; Schober et al.[3]
Ethyl maltol	Y	Y	Y/N	Flavoring	Thornburg et al.[2]; Schober et al.[3]
Benzyl alcohol	Y	–	Y	Flavoring	Schober et al.[3]
Menthol	Y	–	Y	Flavoring	Schober et al.[3]
Vanillin	Y	–	Y	Flavoring	Schober et al.[3]
L-Limonene	Y	–	Y	Flavoring	Schober et al.[3]
β-Pinene	Y	–	N	Flavoring	Schober et al.[3]
Butylated hydroxyanisole	Y	Y	N	Preservative	Thornburg et al.[2]
Butylated hydroxytoluene	Y	Y	N	Preservative	Thornburg et al.[2]
Blue No. 1	Y	N	N	Food coloring	Thornburg et al.[2]

[a]Thornburg et al. measured particle phase composition. Schober et al. did not.

a portion of the ingredients in the particle phase. The limited data available indicate that the food preservatives BHA and BHT are only present in the particle phase.

MEASUREMENT METHODS

Methods for measuring the concentration and composition of e-Cigarette vapors, both particles and gases, fall into two categories. Real-time methods provide temporally resolved data useful for tracking the evolution of the vapor's physical and chemical properties over time. Integrated methods provide a time-averaged measurement of vapor properties, which is useful for quantifying trace components with concentrations below the detection limit of real-time instrumentation. Different techniques are used for the aerosol and gas phase measurements. Not all of these methods are applicable for measuring aerosol and gas phase constituents, characterizing chemical and physical changes that occur in a simulated user's respiratory system, and potential secondhand exposure. This section summarizes available methods and operating techniques. There are numerous books and journal articles that provide detailed theoretical and practical information about these methods.

Aerosol Instrumentation

Aerosol measurement methods that quantify the particle concentration, size distribution, and chemical composition are available. Certain methods can measure one or more of these characteristics either in real time or as a time-integrated average. Particle concentration can be expressed as a number per unit volume or a mass per unit volume. Although either number or mass concentration is the principle measurement, one can be calculated from the other if the particle diameter and density are known or assumed.

Real-Time Aerosol Instruments

These are used frequently to characterize the particle concentration and size distributions produced by e-Cigarettes.[4–6] Table 3.2 describes the various methods. These devices are preferred because they provide a large quantity of data quickly, temporal resolution being on the order of seconds. High temporal resolution is useful for understanding how the particle physical characteristics produced by a device or e-liquid evolve over time. Another advantage is that most types of

Table 3.2 Available Methods to Measure the Real-Time Aerosol Concentration or Size Distribution

Methods	Purpose	Notes on Use
Differential Mobility Analyzers	Determine size distribution and number concentration of particles with diameters from ~ 10 to ~ 1000 nm. Operating principle is the electrical mobility of the particle	Particle evaporation can be significant under certain operating conditions
		E-cig aerosol may need to be diluted to avoid saturating the device
Optical Particle Counters	Measure the number concentration and size distribution of particles with diameters from ~ 0.1 to ~ 20 µm. Operating principle is the intensity of light scattered by a particle illuminated by a laser	Use the density of PG, VG, or a PG:VG mix for the most accurate conversion from number to mass concentration
		E-cig aerosol may need to be diluted to avoid saturating the device
Aerodynamic Particle Sizers	Measure size distribution and mass concentration of particles with diameters from ~ 0.5 to ~ 20 µm. Operating principle is the time for a particle under accelerating flow to traverse a distance between two lasers	E-cig aerosol may need to be diluted to avoid saturating the device
Electric Low Pressure Impactors	Measure size distribution and mass concentration data between 30 nm and 10 µm. The charged particles that deposit on each stage alter the electrical current, which is proportional to the particle mass	Particle evaporation may occur, especially at the smaller diameter stages
Nephelometers	Uses light scattering to measure the mass concentration of all particles within a defined space of the device. Smallest particle diameter detected is ~ 0.3 µm	Calibrate the nephelometer response with PG, VG, or PG:VG mix to obtain the most accurate concentration data
Aerosol Speciation Systems	Measure the mass concentration and chemical composition of particles. Most useful for hydrocarbon like aerosol and oxygenated organic aerosol speciation	The organic fraction speciation is not automatic. It will require advanced analysis to separate and quantify the different species

real-time instruments generate data that spans a wide range of sizes classified into many discrete bins to provide a continuous measure of the size distribution. If these instruments are used properly, they also yield accurate and precise data. Proper use means the user must be aware of potential measurement artifacts.

A potential artifact unique to e-Cigarette particles results from their highly volatile nature. Devices that operate under vacuum or high flow conditions may cause the liquid particles to evaporate during measurement, reducing the measured size distribution. The magnitude of the error depends on the volatility, surface tension and relative composition of the semivolatiles in the aerosol particles. For example, differential mobility analyzer (DMA) measurements can have a negative bias of 10–35% under certain operating conditions.[7]

Given the volatility, high concentration, and short puff duration of e-Cigarette aerosol, researchers may be tempted to set their DMA to a low sheath flow and fast scan rate in order to measure the complete size distribution from a single puff. However, these DMA operating conditions promote evaporation and will maximize the negative bias in the measured particle size distribution.[7] A simple procedure to minimize the evaporation within a DMA is to take advantage of the fact that DMAs internally recycle the air flow (called sheath air) required for proper operation. Measurement of the particle size distribution after sampling multiple puffs to establish a steady-state PG:VG vapor concentration in the sheath air will minimize the negative bias in the size distribution.

Many instruments also are unable to accurately measure the extremely high particle concentrations emitted directly from an e-Cigarette. An accurate measurement requires careful dilution of the aerosol to reduce the concentration without promoting evaporation. A properly operated commercially available or customized dilution system will minimize evaporation by establishing a steady-state PG:VG vapor concentration within the diluter.

Integrated Aerosol Instruments
These collect the particles on a substrate or collection media for subsequent mass or chemical analysis. A variety of particle collection media are available. The most common are glass fiber filters, membrane filters, and polyurethane foam (PUF). CORESTA Method No. 81 and Health Canada Test Method T-115 specify glass fiber substrates because the methods were originally developed for conventional tobacco smoke collection in a laboratory system. Membrane filters are available in various materials for laboratory or field studies. A commonly used material for aerosol sampling is polytetrafluorethylene (PTFE) because it is chemically inert (i.e., will not react with the collected particles), hydrophobic, and not subject to accumulation of electrostatic charge. PUF is not commonly used for general aerosol measurements. However, the high mass fraction of organic aerosol in e-Cigarette vapors increases the versatility of PUF, because the total mass and the speciated organic mass can be measured as described in Gas Phase Methods.

The sampler can capture all particles on a single substrate or use impaction to inertially separate the particles by their size for collection on one or more substrates. There are numerous single stage filter holders available that collect all the particles or have an impaction stage to collect a subfraction of the aerosol.[8] The CORESTA methods, for example, collect all particles. Alternatively, a $PM_{2.5}$ sampler will collect essentially 100% of all particles with aerodynamic diameters smaller than 2.5 μm. A cascade impactor has multiple stages that collect all particles between the lower and upper aerodynamic sizes. The number of size separation stages determines the resolution of the measured mass size distribution.

Like the real-time aerosol instruments, the integrated devices are subject to sample collection artifacts. Again, the most common artifact when sampling e-Cigarette emissions is evaporation of the volatile components. The air velocity through the filter, referred to as the face velocity, drives the evaporation of the volatile components. A higher face velocity promotes evaporation. Face velocities greater than 10 cm/second cause significant evaporative losses.[9] For comparison, CORESTA Recommended Method 81 specifies a face velocity of 0.2 cm/second to minimize evaporative losses. Similarly, the pressure drop across cascade impactor stages, especially stages designed to collect submicrometer particles, can cause evaporation.[10] If evaporative losses are suspected, a PUF filter or sorbent tube placed downstream from the filter is recommended to capture the gases.

The most common analytical method performed on filter substrates is gravimetric analysis using a high precision microbalance (±1 μg) to determine the total mass of the particles collected. Because of the volatile nature of e-Cigarette aerosols, storage of collected filters in a 4°C refrigerator is recommended. Also, because the aerosols contain a high water fraction, equilibration of the filters for 24 hours at 23°C and 35% relative humidity (RH) prior to preweight and postweight measurements will account for water absorption on the filter media.

Other analytical methods can quantify metal and organic species contents. The methods selected will depend on whether or not multiple analyses are desired and the desired minimum detection limit (MDL). Metals can be quantified by x-ray fluorescence (nondestructive), atomic absorption (destructive), or inductively coupled mass spectrometry (destructive). The organic species, including nicotine and flavors, can be chemically extracted from the filter substrate for analysis by gas chromatography and liquid chromatography.

Gas Phase Methods

The measurement of gas phase chemicals emitted by e-Cigarettes has focused on organic compounds, including PG:VG, nicotine, carbonyls, and polycyclic aromatic hydrocarbons. To date, no one has determined whether inorganic compounds, such as inorganic acids, are present in the gas phase. Therefore, this section will focus on methods for sampling organic gases. Gas phase measurement methods quantify the gas concentration expressed as $\mu g/m^3$, ppm, or similar units. Like the aerosol methods, real-time and integrated organic gas measurement methods are available.

Real-Time Organic Gas Instruments

Instruments for the continuous measurement of organic gas concentrations are commercially available. Devices that use photoionization detectors (PIDs) or metal oxide sensors (MOSs) are most conducive to use in e-Cigarette vapor exposure research in the laboratory or public settings. These instruments are small, easy to use, and inexpensive compared to other types of devices. However, both PIDs and MOSs have their disadvantages.

PIDs have excellent linear response over a broad concentration range. This improves the accuracy of the measurement. The primary limitation is that the PID provides a nonselective measurement. It measures the total organic gas concentration without identifying individual species. If used in a study where the organic gases emitted by the e-Cigarette will be diluted significantly, the PID could also have detection limit issues. PIDs typically have a MDL in the ppm range or roughly less than $3 \mu g/m^3$, depending on the gas composition. Given that the organic gas will be primarily PG:VG, the data provided will be a presence—absence assessment of e-Cigarette vapors and not a quantification of the trace organic flavorings or thermal decomposition by-products of potential concern.

MOSs are an established and inexpensive technology for measuring organic gas concentrations with a degree of specificity. However, the current state of the technology limits their use for e-Cigarette vapor exposure research. Their applicability is limited by two factors. One is their poor sensitivity, with detection limits greater than 10 ppm. The other is their positive response to water vapor. However, MOS research is an active area. Researchers are advancing the technology by adding nanostructures and advanced polymers to improve the sensitivity and reduce artifacts that bias the measurements.[11,12]

Highly sensitive real-time formaldehyde gas monitors are available and have been used in e-Cigarette research.[13] These instruments are limited to laboratory research because of their need for chemical reagents, waste disposal, and highly trained operators.

Integrated Organic Gas Methods

Sorbent media are commonly used to collect gas phase samples, especially for organic gases emitted by e-Cigarettes.[14] The type of sorbent selected will depend on the type of gas to be collected. Activated charcoal and organic porous polymer sorbents are commonly used for alkane and polycyclic aromatic hydrocarbon collection. The collected samples are chemically or thermally desorbed and analyzed by gas chromatography–mass spectrometry to identify and quantify (by mass) each species collected. Carbonyls, such as formaldehyde, are collected on 2,4-dintorphyenyhydrazine (DNPH) sorbent. The DNPH sorbent derivatizes the carbonyls for quantitative measurement of the mass by high-pressure liquid chromatography.

Gas samples can be actively drawn onto the sorbent with a pump or passively sampled by diffusion. The selection of active versus passive sampling depends on the objectives of the research. Active sampling will collect more mass on the sorbent, thereby increasing the likelihood of collecting a sample above the MDL. This approach is good for laboratory experiments to measure user and secondhand exposure. Passive badges are small, lightweight devices easily worn by study participants. This alternative is a good strategy for panel studies where a large number of study participants will be recruited.

USER EXPOSURE

This section discusses the physical and chemical changes that occur when a user inhales the emissions from an e-Cigarette. This is important for understanding a user's dose as well as the exhaled fraction that is the source of secondhand exposure. Particle growth within the respiratory system influences particle deposition and gas absorption. These are not only directly related to the nicotine dose received by the user but also the dose of other constituents, such as metals, flavorings, and byproducts produced by the vaping process. The exhaled emissions that consist of particles not deposited or gases not absorbed by the respiratory system are the sole source for potential secondhand exposure.

Evolution of Vapors in the Respiratory System

The evolution of e-Cigarette vapors in a user's respiratory system is directly related to the dose received by the user and the exhaled fraction that leads to secondhand exposure. Previous research with conventional cigarettes has shown the importance of considering the RH and temperature within the airway when predicting particle deposition.[15] The resulting size distribution of the e-Cigarette aerosol in published studies is similar to that reported from conventional cigarettes.[16-18] PG and VG are known humectants, a substance that has an affinity for water. As common components of e-Cigarette vapors, PG and VG drive the physical and chemical changes of the emissions inhaled by the user. Within the humid environment of the respiratory tract, the PG and VG will promote the heterogeneous growth of the particles as water is absorbed. Water absorption can also affect the gas–particle partitioning of semivolatile species. This could drive water-soluble species, like nicotine, into the aerosol phase. The altered aerosol size distribution will affect the fraction of the particles deposited in the lungs and where deposition occurs. Therefore, understanding how the aerosol size distribution changes is important in order to accurately model lung deposition.

Particle Growth

RTI International developed an e-Cigarette emissions generation/ sampling system to mimic the physiologically relevant temperature and humidity found in a user's respiratory system and thereby study the growth of e-Cigarette aerosols in a user's respiratory tract (Fig. 3.1). The system mixed the vapors with particle-free room air heated to 37°C and humidified to 95%. The simulated lung was designed to mimic the flow, average velocity, residence time, and environmental conditions of an adult or teenager inhaling e-Cigarette vapors. A vacuum pump connected to the bottom of the 3-foot-long, 1-inch-diameter PVC tube simulated the lung. Controllable parameters include puff interval, puff duration, inhalation volume, pause (time between inhalation and exhalation), and residence time. This study used a puff duration of 2 seconds and total volume of 66.6 mL every 75 seconds. Solenoid valves allow the flow rate through the conditioning tube to be adjusted to a residence time, set at 20 seconds for this research, mimicking the human airway.[19] Actual activation of the heating coil within the e-Cigarette was initiated by the system operator. An SMPS–CPC system (scanning mobility particle

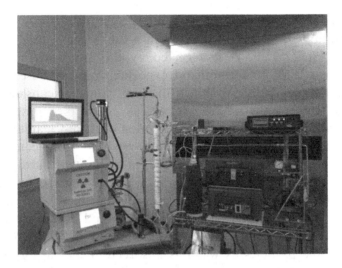

Figure 3.1 An apparatus to simulate the physiological conditions within the human respiratory tract to measure the evolution of e-Cigarette vapors and obtain an accurate measurement of the particle size distribution and gas concentration inhaled by the user.

sizer–condensation particle counter; TSI Inc., Shoreview, MN) operating at steady-state conditions to minimize evaporative loss (see previous section) measured the aerosol size distribution. A single, commercially available refillable cartomizer style device was used. Prior to each test, the battery was fully charged and the device was filled with either a liquid designed to recreate the flavor of a conventional cigarette (Liquid 1) or a liquid flavored as fruit punch (Liquid 2). Both liquids had a nicotine concentration of 0.5 mg/mL and identical ingredients listed as PG, VG, and natural and/or artificial flavorings.

We found that the warm, humid environment in the human airways induces growth of a smaller particle mode (Fig. 3.2). This smaller mode is presumably due to water-induced activation of condensation nuclei. The hygroscopic properties of pure PG in humid nonconden-sing environments has previously been shown to increase aerosol diameter up to ∼40%.[20] A major difference between this work and that of Petters et al. is that it did not include the temperature variable, thereby eliminating evaporation as a competing mass transfer pathway. Under dry conditions (RH < 10%), aerosol from Liquid 1 was log-normally distributed with a CMD of 218 nm, while aerosol from Liquid 2 was log-normally distributed centered at 333 nm with a noticeable shoulder appearing at approximately 400 nm (Fig. 3.2). When emissions from

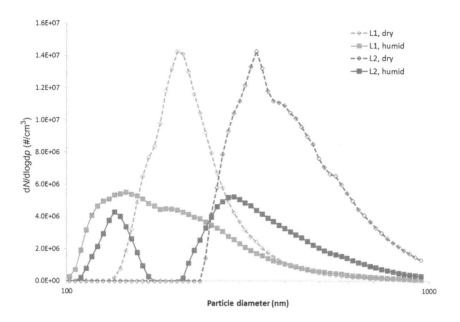

Figure 3.2 Aerosol size distribution of two e-liquids (L1, L2) measured under dry and humid conditions using RTI's simulated respiratory tract.

the e-Cigarette were exposed to humid conditions (RH = 95%), the larger mode remained though concentrations were reduced relative to dry conditions, but a second smaller mode appeared at 145 nm for Liquid 1 and 135 nm for Liquid 2.

Recent data support heterogeneous aerosol growth via water-induced activation of condensation nuclei. Mikheev et al.[21] measured the size distribution of particles by direct injection into a DMA. They found a bimodal distribution, with the smaller mode at 11−25 nm produced by metals released by the device heater coil. The zinc, chromium, and other transition metals identified are the source of the condensation nuclei required for heterogeneous nucleation and condensation. This mechanism supports the bimodal distribution measured under steady-state conditions by Zhang et al.[22] Their primary and secondary size distribution peaks for pure PG and VG occurred at approximately 400 and 150 nm, respectively. Their size distribution data and supposition that the bimodal distribution resulted from the heterogeneous condensation of vapors are in agreement with the RTI size distribution data and conclusion.

The physical and chemical mechanisms that drive the heterogeneous growth of chemically complex e-Cigarette particles in the user's respiratory system are poorly understood. Aerosol science theory explains aerosol growth in a two-phase system,[23] such as a PG:water mixture. However, the different size distributions that RTI measured for two liquids where PG and water are the primary constituents cannot be explained. The differences in the minor constituents (flavorings, preservatives, artificial colors) between the two liquids are a likely cause. As noted earlier in this chapter, these minor constituents are semivolatile. Therefore, the compounds will partition between the gas and particle phases to achieve equilibrium.[24] Given that conventional tobacco smoke contains nicotine and other semivolatile organic compounds,[25,26] it is likely that the variability in types and quantities of flavors and other minor constituents in e-liquids will affect the evolution of the vapors inside a user's respiratory tract. The magnitude of these changes, their importance on respiratory deposition, variability within different e-liquids, and specific device settings need to be experimentally determined. To allow comparison between these expected studies, a reference e-liquid and device should be produced to serve as a control, given the number of researchers expected to perform these experiments using a variety of e-liquids, devices, and measurement methods.[2]

Deposition in the Respiratory System

Particle deposition in the respiratory system is frequently modeled. Two advanced and widely used models are the International Commission on Radiological Protection (ICRP)[27] and the Multiple Path Particle Dosimetry (MPPD).[28] They estimate regional and total deposition across the range of particle sizes relevant to e-Cigarettes and breathing conditions for adults and children. Both models have been used to estimate deposition in a user's airways.

Two studies that can be easily compared are Zhang et al.[22] and Thornburg et al.[2] Zhang et al. modeled deposition using the ICRP model applied to an adult worker with a light average ventilation rate of 1.2 m³/hour and a heavy rate of 1.69 m³/hour. The model predicted total particle deposition in the respiratory tract from 20% to 27%, with 9% to 18% deposited in the alveolar region. Thornburg et al. used the MPPD model to simulate e-Cigarette aerosol deposition in the respiratory system of a teenage boy. The model predicted total particle deposition of 47%, with 40% in the alveolar region. These modeled deposition fractions are significantly lower than the only currently available experimentally measured values. St. Helen et al.[29] experimentally found that 84.4–91.7% of the inhaled aerosol was retained by 13 healthy adult e-Cigarette users.

The differences between the modeled and experimental data can be attributed to four potential causes.[30] Hygroscopic growth caused by the warm, humid environment within the respiratory tract was discussed earlier. Zhang et al. correctly noted that their predicted deposition results were biased low because their experimental system did not account for particle growth by heterogeneous condensation (as noted above and by Longest and Xi[31]). Their experimental system produced a dry e-Cigarette aerosol for the measurement of the size distribution used as input to the ICRP model. A second factor not considered by either Zhang et al. or Thornburg et al. was the "cloud effect" that increases deposition fractions at very high aerosol concentrations.[32,33] At particle concentrations greater than 10^5 per cm^3, the particle–particle interactions decrease the effective drag on the particles, thereby increasing their mobility and promoting more deposition in the upper tracheobronchial region.[34] The other two mechanisms that could increase particle deposition in the upper tracheobronchial region are particle coagulation and electrical charge.[30] These mechanisms increase the particle mobility via inertial (coagulation) and image (electrical charge) forces. The relative importance of these mechanisms for e-Cigarette aerosol deposition in airways is unknown and needs further investigation.

SECONDHAND AND TERTIARY EXPOSURES

Secondhand exposure is the inhalation of the aerosol and gases by a nonuser. Tertiary exposure occurs when nicotine and other semivolatile species deposited or absorbed onto surfaces are released back into the air, thereby producing another inhalation exposure or are transferred to a person's skin producing a dermal exposure. Many people perceive secondhand exposure to e-Cigarette vapors to be harmful to health.[35] The potential for secondhand and tertiary exposures to vapor constituents is determined solely by the aerosol and gases exhaled by the user. e-Cigarettes, unlike conventional cigarettes or other combustible tobacco products, do not produce sidestream emissions (e.g., the smoke produced by tobacco combustion in between puffs).

The volatile nature of the aerosol constituents is a critical factor that determines secondhand and tertiary exposures. The partitioning between the gas and particle phases for each chemical species in the exhaled vapors will change over time as the particles evaporate. The evaporation determines the chemical species mass concentrations in the gas and particle phases, thereby determining the relative deposition and absorption efficiency within the respiratory system. The same gas–particle partitioning dynamics also determine the mass of particles deposited or gases absorbed onto surfaces that contribute to tertiary exposures. Vapor dispersion, transport, and removal mechanisms also determine the secondhand and tertiary exposure concentrations. The magnitude of these exposures lead to the internal dose that will determine whether or not adverse toxicological outcomes will occur.

Aerosol Evaporation, Dispersion, and Transport

The evaporation of the volatile components and resulting change in the size of particles can be estimated by established aerosol theory.[1] The critical parameters in Eq. (3.1) are the partial vapor pressures of individual chemicals (P), the initial particle size (d_p), and the diffusion coefficient ($D_{a,b}$). The partial vapor pressures determine the gas–particle partitioning. The initial particle diameter determines the particle surface area to volume ratio. The diffusion coefficient specifies the mass transfer rate from the particle to the atmosphere.

This equation provides a theoretical description of the visual changes to the vapor exhaled by a user. Soon after exhalation, the particle evaporation rate is low because the partial pressures far from (P_∞) and near the particle surface (P_d) are similar. As the particles disperse in the environment, P_∞ will decrease, thereby increasing the rate of evaporation. The evaporation rate will further increase as d_p decreases as the surface area available for evaporation relative to the particle volume continues to increase. Eventually, the aerosol evaporation will cease and an equilibrium size distribution will be maintained. Controlled, in-depth experiments to characterize e-Cigarette aerosol evaporation have not been conducted to prove this theory. Blair et al.[36] however, present particle and VOC concentration and particle size distribution data that strongly suggest that evaporation of e-Cigarette aerosol cannot be neglected when assessing secondhand exposure.

$$d_{p,f} = \sqrt{2 \frac{\left(\frac{\rho_a R}{4\mathscr{D}_{a,b}M_a\left(\frac{P_\infty}{T_\infty} - \frac{P_d}{T_d}\right)}\frac{d_p^2}{2} - \Delta t\right)}{\frac{\rho_a R}{4\mathscr{D}_{a,b}M_a\left(\frac{P_\infty}{T_\infty} - \frac{P_d}{T_d}\right)}}} \qquad (3.1)$$

The dispersion and transport of the aerosol cloud into a space influences secondary exposure by diluting the particles and gases to a concentration orders of magnitude lower than the exhaled concentration. Outdoors, where the volume of the space is essentially infinite, the potential for secondhand and tertiary exposures is low unless the subjects are within a few feet of the user. Therefore, the remainder of this discussion will focus on indoor environments where the volume of the space is constrained.

Two chamber studies provide empirical evidence that significant aerosol evaporation in addition to dispersion and removal by the ventilation system occurs. The study by Schripp et al.[13] had a single participant in an 8-m^3 chamber operating at 24°C and 44%RH. The location of sample collection and whether or not the chamber was thoroughly mixed were not reported. They found carbonyl concentrations ranging from 2 μg/m^3 to 25 μg/m^3, depending on species. However, all other gas phase constituents, including PG, were below MDL. The fact that PG was not detected is not understood. The maximum particle concentration reported was approximately 3E+04 per cm^3 about 7 minutes into the experiment. Extrapolating the Schripp et al. data from their Figure 4a to get a median diameter of 35 nm, and assuming the particle density of 1 g/cm^3 (water and PG), the number concentration can be converted to a mass concentration of 0.6 μg/m^3. Given the high carbonyl concentration, low particle concentration, and small particle size distribution, it is highly likely that significant evaporation occurred. RTI conducted a similar chamber study with a single user standing in the corner of a 30-m^3 chamber operating at 23°C, 40%RH, and 0.5 air changes per hour. The subject used their personal device and e-liquid. PM$_{2.5}$ mass concentration was measured with nephelometers (MicroPEM v3.2a, RTI International) worn by the user and also placed 1 m and 2 m away on a table 1 m above the floor. The PM$_{2.5}$ concentration spikes at the source identify when puffs occurred, and the subsequent lower concentration spikes demonstrate the time lag for detectable amounts of

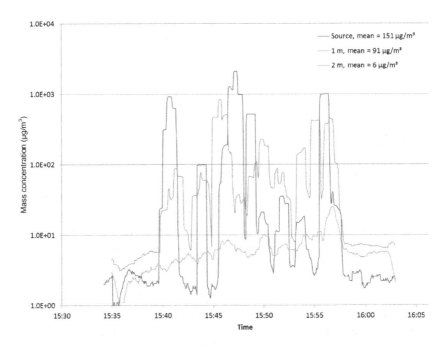

Figure 3.3 PM$_{2.5}$ concentration produced by a single e-Cigarette user inside RTI's 30-m^3 chamber. The decrease in the PM$_{2.5}$ concentration as distance from the user increases is evidence of dispersion, capture by the ventilation system, and aerosol evaporation.

the aerosol to transport within the chamber (Fig. 3.3). The reduction in average PM$_{2.5}$ concentration from 151 μg/m^3 at the source to 91 and 6 μg/m^3 at 1 and 2 m, respectively, results from a combination of dispersion within the chamber, removal by the ventilation system, and evaporation of the particles as described earlier. Even with these reductions, a bystander 2 m from the user would be exposed to an average concentration of 6 μg/m^3 with a peak exposure of 25 μg/m^3.

Secondhand Exposure Levels

Secondhand exposure levels are determined by source strength, distribution of the sources within the room, a person's proximity to the sources, room volume, and room ventilation conditions. It is intuitively obvious that the number of e-Cigarette users and their location in a room define the source strength and potential spatial gradients in the vapor concentration.

Reported e-Cigarette particle concentration data inside real buildings are limited. The two existing studies found between 197 μg/m^3 and more than 300 µg/m³.[3,37] When adjusted for the number of users, the $PM_{2.5}$ concentration spread increases to between 4.4 (Soule et al.) and 65.7 (Schober et al.) µg/m³ per user. Schober et al., as part of their thorough evaluation, also reported total PG and VG (particle and gas phases) concentration as 110 µg/m³ to 395 and 59–81 µg/m³, respectively, as well as measurable levels of flavorings such as vanillin, menthol, and benzyl alcohol. Interestingly, Schober et al. did not detect an increase in aldehyde and L-limonene concentrations during e-Cigarette use, probably because of the prevalence of these compounds in furniture and cleaners.

Schober et al. and Ballbe et al.[38] are the two studies that assessed secondhand exposure to nicotine from e-Cigarettes inside occupied buildings. Schober et al. found nicotine concentration from <MDL to 4.6 µg/m³. Ballbe et al. found a mean concentration of 0.13 µg/m³ with a geometric standard deviation of 2.4 in five participants.

Across all three studies, the within- and between-study variabilities in the measured concentrations are most likely due to differences in location of the users in relation to the sources, the ventilation system return/supply air ducts, fraction of ventilation air filtered and recycled, and the influence of open doors or windows during the course of data collection.

Other studies have measured potential secondhand exposure to e-Cigarette particles and gases under controlled laboratory conditions (Table 3.3). Only studies that reported vapor constituent concentrations are included. Studies that presented data on a total mass or per-puff basis are not included because the result cannot be easily extrapolated to meaningful estimates of secondhand exposure. The reported concentrations of all constituents measured are highly variable across the three studies. Differences in the experimental details are the most likely cause. In addition to differences in chamber volume and air exchange rate, all three studies had differences in the chamber ventilation, vapor generation, and sample collection details. For example, Czogala et al.[39] appears to be the only study with mixing fans inside the chamber to uniformly mix the produced vapors. The studies also generated different quantities of vapors because the number of puffs, puff volume, puff duration, and puff interval varied. The variation in sample collection media, flow, and duration also influenced the reported concentrations. For example in the Schripp et al. experiments, the last 5 minutes of sample collection (33% of the total

Table 3.3 Comparison of Laboratory Studies Conducted to Assess Secondhand Exposure to e-Cigarette Vapors

Study	Conditions	Avg PM$_{2.5}$ (μg/m^3)	Nicotine (μg/m^3)	PG:VG (μg/m^3)	Carbonyls (μg/m^3)	Comment
Czogala et al.[39]	39-m^3 chamber. Air changes per hour (ACH) 1.37–12.6, mixing fans, machine and single user sources (5 subjects, 5 e-liquids)	Machine: 13.6 User: 151.7	Machine: 2.53 User: 3.3	Not reported	Not measured	ACH 4× lower during single user experiments
Geiss et al.[40]	30-m^3 chamber, 0.5 ACH, 23°C, 50%RH, machine generated, 2 e-liquids	7E+06 particles/L	0.2–0.6	PG 1400–2200 at peak VG 60–136 at peak	< MDL	
Schripp et al.[13]	8-m^3 chamber, 0.3 ACH, 24°C, 44%RH, single user (3 subjects, 3 e-liquids)	0.6	Not reported	PG <MDL	Acetone: 17–25 Formaldehyde: 8–16 Acetaldehyde: 2–3	Calculated PM$_{2.5}$ mass concentration as described previously

sample volume) collected only 10% of the VOC mass. The result is VOC mass concentrations that are biased low.

Potential for Tertiary Exposure

The available data on tertiary exposure to nicotine released by e-Cigarettes are limited to a single study. Goniewicz and Lee[41] found a measurable increase in nicotine load on five common indoor surfaces. Flooring and glass had significant increases ($P < 0.05$). The findings are consistent with studies of tertiary exposure to nicotine from conventional cigarettes.[42] The potential for tertiary exposure to other e-Cigarette constituents, such as flavorings, remains to be investigated.

CONCLUSION

Given the differences in aerosol size and composition reported in the literature, evaluation of the potential risks associated with e-Cigarette use requires a more cohesive and detailed approach. The current literature uses a variety of e-Cigarette devices and liquids, which hinders comparison of results. The aerosol dynamics that drive particle deposition in the respiratory system are poorly understood. Experimental systems need to be developed to characterize this phenomenon and enable the accurate lung deposition data necessary to determine the dose that could lead to toxicological effects. Then experiments can be designed to accurately measure the aerosol size distribution and concentration for inputs into the ICRP or MPPD models. Accurate modeling data will clarify the role of user physiology, e-liquid type, and device characteristics that may influence the variability in nicotine dose observed.[29] Additionally, since secondhand exposure is caused only by the fraction exhaled by the user, accurate respiratory deposition model outputs will provide information necessary to better understand the potential magnitude of secondhand and tertiary exposures. These data will inform the design of laboratory and panel exposure studies to quantify the risk to bystanders. Lastly, there is need for a standardized e-Cigarette device, liquid, and testing protocol in order to generate a body of data to develop a realistic picture of the user dose and secondhand exposure. This approach will be similar to that taken with the adoption of the US Federal Trade Commission method in 1966 and development of reference cigarettes in 1969.[43,44] The need for such an effort for e-Cigarettes has recently been advocated.[45] Ultimately, with standardized devices and protocols to provide representative baseline data, the research community can conduct risk assessments of e-Cigarettes needed to inform future policy decisions and manufacturer product specifications.

REFERENCES

1. Hinds WC. *Aerosol sci technol: properties, behavior, and measurement of airborne particles.* 2nd ed. New York: John Wiley & Sons; 1999<http://www.wiley.com/WileyCDA/WileyTitle/productCd-0471194107.html>.

2. Thornburg J, Malloy Q, Cho S-H, Studabaker W, Lee Y. *Exhaled electronic cigarette emissions: what's your secondhand exposure?* RTI Press; 2015<http://www.rti.org/publications/abstract.cfm?pubid=24019>.

3. Schober W, Szendrei K, Matzen W, et al. Use of electronic cigarettes (e-cigarettes) impairs indoor air quality and increases FeNO levels of e-cigarette consumers. *Int J Hyg Environ Health* 2014;**217**(6):628–37. Available from: http://dx.doi.org/10.1016/j.ijheh.2013.11.003.

4. McAuley TR, Hopke PK, Zhao J, Babaian S. Comparison of the effects of e-cigarette vapor and cigarette smoke on indoor air quality. *Inhal Toxicol* 2012;**24**(12):850–7. Available from: http://dx.doi.org/10.3109/08958378.2012.724728.

5. Ingebrethsen BJ, Cole SK, Alderman SL. Electronic cigarette aerosol particle size distribution measurements. *Inhal Toxicol* 2012;**24**(14):976–84. Available from: http://dx.doi.org/10.3109/08958378.2012.744781.

6. Fuoco FC, Buonanno G, Stabile L, Vigo P. Influential parameters on particle concentration and size distribution in the mainstream of e-cigarettes. *Environ Pollut* 2014;**184**:523–9. Available from: http://dx.doi.org/10.1016/j.envpol.2013.10.010.

7. Khlystov A. Effect of aerosol volatility on the sizing accuracy of differential mobility analyzers. *Aerosol Sci Technol* 2014;**48**(6):604–19. Available from: http://dx.doi.org/10.1080/02786826.2014.899681.

8. Lee KW, Mukund R. Filter collection. In: Baron PA, Willeke K, editors. *Aerosol measurement: principles, techniques, and applications*. 2nd ed. John Wiley & Sons, Inc.; 2001<http://onlinelibrary.wiley.com/doi/10.1002/9781118001684.ch7/summary>.

9. Hering SV, Appel BR, Cheng W, et al. Comparison of sampling methods for carbonaceous aerosols in ambient air. *Aerosol Sci Technol* 1990;**12**(1):200–13. Available from: http://dx.doi.org/10.1080/02786829008959340.

10. Zhang XQ, McMurry PH. Theoretical analysis of evaporative losses from impactor and filter deposits. *Atmos Environ (1967)* 1987;**21**(8):1779–89. Available from: http://dx.doi.org/10.1016/0004-6981(87)90118-1.

11. Lange D, Hagleitner C, Hierlemann A, Brand O, et al. Complementary metal oxide semiconductor cantilever arrays on a single chip: mass-sensitive detection of volatile organic compounds. *Anal Chem* 2002;**74**(13):3084–95. Available from: http://dx.doi.org/10.1021/ac011269j.

12. Wolfrum EJ, Meglen RM, Peterson D, et al. Metal oxide sensor arrays for the detection, differentiation, and quantification of volatile organic compounds at sub-parts-per-million concentration levels. *Sens Actuat B Chem* 2006;**74**(1):322–9. Available from: http://dx.doi.org/10.1016/j.snb.2005.09.026.

13. Schripp T, Markewitz D, Uhde E, Salthammer T. Does e-cigarette consumption cause passive vaping? *Indoor Air* 2012;(1). Available from: http://dx.doi.org/10.1111/j.1600-0668.2012.00792.x.

14. Goniewicz ML, Knysak J, Gawron M, et al. Levels of selected carcinogens and toxicants in vapour from electronic cigarettes. *Tob Control* 2014;**23**(2):133–9. Available from: http://dx.doi.org/10.1136/tobaccocontrol-2012-050859.

15. Ingebrethsen BJ, Alderman SL, Ademe B. Coagulation of mainstream cigarette smoke in the mouth during puffing and inhalation. *Aerosol Sci Technol* 2011;**45**(12):1422–8. Available from: http://dx.doi.org/10.1080/02786826.2011.596863.

16. McGrath C, Warren N, Biggs P, McAughey J. Real-time measurement of inhaled and exhaled cigarette smoke: implications for dose. *J Phys Conf Series.* 2009;**151**:12018. Available from: http://dx.doi.org/10.1088/1742-6596/151/1/012018.

17. McAughey J, Adam T, McGrath C, Mocker C, Zimmermann R. Simultaneous on-line size and chemical analysis of gas phase and particulate phase of mainstream tobacco smoke. *J Phys Conf Ser* 2009;**151**:12017. Available from: http://dx.doi.org/10.1088/1742-6596/151/1/012017.

18. Peeler CL. Cigarette testing and the federal trade commission: a historical overview. In: Libbey J, ed. *Monograh 7: The FTC cigarette test method for determining tar, nicotine, and carbon monoxide yields of U.S. cigarettes*, 1994:1–8. http://cancercontrol.cancer.gov/brp/TCRB/monographs/7/ [accessed 07.08.15].

19. Sahu SK, Tiwari M, Bhangare RC, Pandit GG. Particle size distribution of mainstream and exhaled cigarette smoke and predictive deposition in human respiratory tract. *Aerosol Air Qual Res* 2013. Available from: http://dx.doi.org/10.4209/aaqr.2012.02.0041.

20. Petters MD, Kreidenweis SM, Snider JR, Koehler KA, Wang Q, Prenni AJ, et al. Cloud droplet activation of polymerized organic aerosol. *Tellus B* 2006. Available from: http://dx. doi.org/10.1111/j.1600-0889.2006.00181.

21. Mikheev VB, Brinkman MC, Granville CA, Gordon SM, Clark PI. Real-time measurement of electronic cigarette aerosol size distribution and metals content analysis. *Nicotine Tob Res* 2016;**18**(9):1895–902. Available from: http://dx.doi.org/10.1093/ntr/ntw128.

22. Zhang Y, Sumner W, Chen D-R. In vitro particle size distributions in electronic and conventional cigarette aerosols suggest comparable deposition patterns. *Nicotine Tob Res* 2013;**15**(2):501–8. Available from: http://dx.doi.org/10.1093/ntr/nts165.

23. Tang IN, Munkelwitz HR. Aerosol growth studies—III ammonium bisulfate aerosols in a moist atmosphere. *J Aerosol Sci* 1977;**8**(5):321–30. Available from: http://dx.doi.org/10.1016/0021-8502(77)90019-2.

24. Pankow JF. An absorption model of gas/particle partitioning of organic compounds in the atmosphere. *Atmos Environ* 1994;**28**(2):185–8. Available from: http://dx.doi.org/10.1016/1352-2310(94)90093-0.

25. Liang C, Pankow JF. Gas/particle partitioning of organic compounds to environmental tobacco smoke: partition coefficient measurements by desorption and comparison to urban particulate material. *Environ Sci Technol* 1996;**30**(9):2800–5. Available from: http://dx.doi.org/10.1021/es960050x.

26. Pankow JF. A consideration of the role of gas/particle partitioning in the deposition of nicotine and other tobacco smoke compounds in the respiratory tract. *Chem Res Toxicol* 2001;**14**(11):1465–81. Available from: http://dx.doi.org/10.1021/tx0100901.

27. ICRP. ICRP Publication 66: *Human respiratory Tract Model for Radiological Protection* (No. 66). Elsevier Health Sciences; 1994.

28. Asgharian B, Price O, Miller F, Subramaniam R, Cassee FR, Freijer J, et al. Multiple-path particle dosimetry model (MPPD v 2.11): a model for human and rat airway particle dosimetry. Hamner Institutes for Health Sciences Applied Research Associates (ARA), National Institute for Public Health and the Environment (RIVM), and Ministry of Housing, Spatial Planning and the Environment, Editor; 2009.

29. St. Helen G, Havel C, Dempsey DA, Jacob P, Benowitz NL. Nicotine delivery, retention and pharmacokinetics from various electronic cigarettes. *Addiction* 2016;**111**(3):535–44. Available from: http://dx.doi.org/10.1111/add.13183.

30. Hofmann W, Morawska L, Bergmann R. Environmental tobacco smoke deposition in the human respiratory tract: differences between experimental and theoretical approaches. *J Aerosol Med* 2001;**14**(3):317–26. Available from: http://dx.doi.org/10.1089/089426801316970277.

31. Longest PW, Xi J. Condensational growth may contribute to the enhanced deposition of cigarette smoke particles in the upper respiratory tract. *Aerosol Sci Technol* 2008;**42**(8):579–602. Available from: http://dx.doi.org/10.1080/02786820802232964.

32. Martonen TB. Deposition patterns of cigarette smoke in human airways. *Am Ind Hyg Assoc J* 1992;**53**(1):6–18. Available from: http://dx.doi.org/10.1080/15298669291359249.

33. Martonen TB, Musante CJ. Importance of cloud motion on cigarette smoke deposition in lung airways. *Inhal Toxicol* 2000;**12**(suppl 4):261–80. Available from: http://dx.doi.org/10.1080/08958370050165120.

34. Broday DM, Robinson R. Application of cloud dynamics to dosimetry of cigarette smoke particles in the lungs. *Aerosol Sci Technol* 2003;**37**(6):510–27. Available from: http://dx.doi.org/10.1080/02786820300969.

35. Mello S, Bigman CA, Sanders-Jackson A, Tan ASL. Perceived harm of secondhand electronic cigarette vapors and policy support to restrict public vaping: results from a national

survey of US adults. *Nicotine Tob Res* October 2015;ntv232. Available from: http://dx.doi. org/10.1093/ntr/ntv232.

36. Blair SL, Epstein SA, Nizkorodov SA, Staimer N. A real-time fast-flow tube study of VOC and particulate emissions from electronic, potentially reduced-harm, conventional, and reference cigarettes. *Aerosol Sci Technol* 2015;**49**(9):816−27. Available from: http://dx.doi.org/ 10.1080/02786826.2015.1076156.

37. Soule EK, Maloney SF, Spindle TR, Rudy AK, Hiler MM, Cobb CO. Electronic cigarette use and indoor air quality in a natural setting. *Tob Control* February 2016. Available from: http://dx.doi.org/10.1136/tobaccocontrol-2015-052772.

38. Ballbè M, Martínez-Sánchez JM, Sureda X, Fu M, Pérez-Ortuño R, Pascual JA, et al. Cigarettes vs. e-cigarettes: passive exposure at home measured by means of airborne marker and biomarkers. *Environ Res* 2014;**2014**. Available from: http://dx.doi.org/10.1016/j. envres.2014.09.005.

39. Czogala J, Goniewicz ML, Fidelus B, Zielinska-Danch W, Travers MJ, Sobczak A. Secondhand exposure to vapors from electronic cigarettes. *Nicotine Tob Res* 2014;**16** (6):655−62. Available from: http://dx.doi.org/10.1093/ntr/ntt203.

40. Geiss O, Bianchi I, Barahona F, Barrero-Moreno J. Characterisation of mainstream and passive vapours emitted by selected electronic cigarettes. *Int J Hyg Environ Health* 2015;**218** (1):169−80. Available from: http://dx.doi.org/10.1016/j.ijheh.2014.10.001.

41. Goniewicz ML, Lee L. Electronic cigarettes are a source of thirdhand exposure to nicotine. *Nicotine Tob Res* 2015;**17**(2):256−8. Available from: http://dx.doi.org/10.1093/ntr/ntu152.

42. Matt GE, Quintana PJE, Destaillats H, et al. Thirdhand tobacco smoke: emerging evidence and arguments for a multidisciplinary research agenda. *Environ Health Perspect* 2011;**119** (9):1218−26. Available from: http://dx.doi.org/10.1289/ehp.1103500.

43. Pillsbury HC. Cigarette testing and the federal trade commission: a historical overview. In: Libbey J, ed. *Monograh 7: the FTC cigarette test method for determining tar, nicotine, and carbon monoxide yields of U.S. cigarettes*; 1994:9−14. http://cancercontrol.cancer.gov/brp/ TCRB/monographs/7/ [accessed 07.08.15].

44. Roemer E, Schramke H, Weiler H, et al. Mainstream smoke chemistry and in vitro and in vivo toxicity of the reference cigarettes 3R4F and 2R4F. *Contrib Tob Res* 2012.(1). Available from: http://dx.doi.org/10.2478/cttr-2013-0912.

45. Brown CJ, Cheng JM. Electronic cigarettes: product characterisation and design considerations. *Tob Control* 2014;**23**(Suppl. 2):ii4−ii10. Available from: http://dx.doi.org/10.1136/ tobaccocontrol-2013-051476.

CHAPTER 4

Biomarkers for Assessment of Chemical Exposures From e-Cigarette Emissions

S.S. Hecht

INTRODUCTION

A workshop held by the U.S. National Institutes of Health entitled "NIH Electronic Cigarette Workshop: Developing a Research Agenda" concluded that "studies on the acute use of e-Cigarettes will require urine or plasma biomarkers that can provide an objective indicator of dose."[1] Biomarkers of exposure, which are usually metabolites of specific toxicants or carcinogens, have the advantage of bypassing many of the uncertainties associated with measurement of emissions from a given tobacco or e-Cigarette product. These uncertainties arise from the use of artificial smoking conditions that might not accurately reflect the ways in which humans actually use the products in question. Thus urine, plasma, salivary, and breath biomarkers, which in most cases are quantifiable metabolites of specific tobacco smoke or e-Cigarette emission constituents, can provide objective and useful information on actual human exposure to specific toxicants or carcinogens. This chapter will review some biomarkers of exposure which have been widely used to monitor human uptake of tobacco smoke constituents and will then present data on the application of these biomarkers in studies of e-Cigarette users.

BIOMARKERS OF EXPOSURE IN CIGARETTE SMOKERS

Carbon monoxide (CO) is formed during the incomplete combustion of tobacco during smoking and is a reliable biomarker of cigarette smoking. Either exhaled CO or carboxyhemoglobin can be quantified. Typical values of exhaled CO were 20–30 ppm for smokers and 4–7 ppm for nonsmokers while carboxyhemoglobin levels ranged from 4% to 7% in smokers and 1% to 2% in nonsmokers.[2] CO is likely

Analytical Assessment of e-Cigarettes. DOI: http://dx.doi.org/10.1016/B978-0-12-811241-0.00004-8

related to complications of atherosclerosis and other cardiovascular diseases in cigarette smokers.[3]

Tobacco alkaloids are the most widely used and specific biomarkers of tobacco product exposure. Nicotine and its metabolites can be quantified in blood or urine. The most reliable and practical method is liquid chromatography-tandem mass spectrometry (LC-MS/MS), which has been applied in many large studies. A panel of urinary nicotine metabolites including unchanged nicotine, cotinine, *trans*-3'-hydroxycotinine and their glucuronides as well as nicotine-*N*-oxide represents 85–90% of the nicotine dose taken in by a smoker. The nicotine metabolic pathways leading to the formation of these compounds are illustrated in Fig. 4.1. CYP2A6, UGT2B10, and FMO3 are the major enzymes involved in nicotine metabolism. In most smokers, the predominant pathway is catalysis of nicotine C-oxidation by CYP2A6 resulting ultimately in cotinine, which is further metabolized to *trans*-3'-hydroxycotinine by CYP2A6. Nicotine, cotinine, and *trans*-3'-hydroxycotinine also undergo glucuronidation, catalyzed by UGT2B10 and UGT2B17 among others. In one recent study, typical values for total nicotine (meaning free nicotine plus its glucuronide), total cotinine, and total *trans*-3'-hydroxycotinine in European Americans who smoked an average of 20 cigarettes per day were 5.42, 10.7, and 16.9 nmol/mL urine with percentages of glucuronidation being 34%, 58%, and 24% for nicotine glucuronide, cotinine glucuronide, and *trans*-3'-hydroxycotinine glucuronide, respectively, while the amount of nicotine-*N*-oxide was 1.76 nmol/mL urine.[4] Other sources of nicotine exposure that are generally comparable to those in cigarette smokers include other tobacco products such as smokeless tobacco, nicotine replacement therapy products such as nicotine gum and nicotine patch, and of course, e-Cigarettes.

Minor tobacco alkaloids include anabasine, anatabine, and nornicotine; the latter is also a minor metabolite of nicotine. Their structures are illustrated in Fig. 4.2. In one recent study, these tobacco-specific compounds were quantified in more than 800 smokers with a wide range of tobacco exposures.[5] The median urinary concentrations of anabasine, anatabine, and nornicotine were 5.53, 4.02, and 98.9 ng/mL urine (0.034, 0.025, and 0.67 nmol/mL urine), respectively. Anabasine and anatabine have been suggested as biomarkers to identify tobacco use among subjects on nicotine replacement therapy, because they are specific to tobacco products.

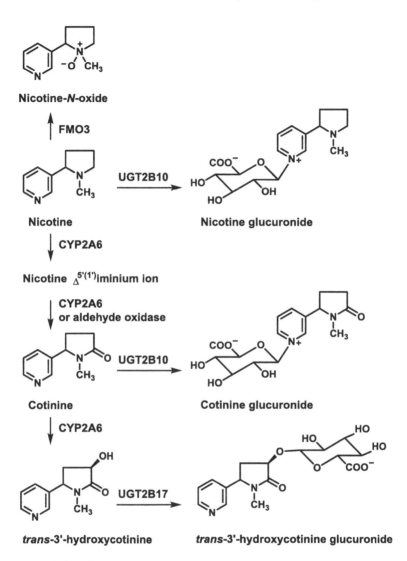

Figure 4.1 Metabolism of nicotine by oxidation and glucuronidation. From Murphy SE, Park S-SL, Thompson EF, Wilkens LR, Patel Y, Stram DO et al. Nicotine N-glucuronidation relative to N-oxidation and C-oxidation and UGT2B10 genotype in five ethnic/racial groups. *Carcinogenesis* 2014;**35**:2526–33.

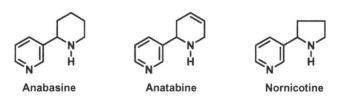

Figure 4.2 Structures of anabasine, anatabine, and nornicotine.

Tobacco-specific nitrosamines are an important class of compounds which also provide biomarkers that are specific to tobacco use. Although there are seven tobacco-specific nitrosamines in tobacco products, two of them—4-(methylnitrosamino)-1-(3-pyridyl)-1-butanone (NNK) and N'-nitrosonornicotine (NNN)—have been extensively studied because of their strong carcinogenic activities.[6,7] NNK and NNN are considered "carcinogenic to humans" by the International Agency for Research on Cancer.[8] The widely used biomarker of NNK exposure is total 4-(methylnitrosamino)-1-(3-pyridyl)-1-butanol (NNAL), comprising NNAL and its N- and O-glucuronides, which are excreted in the urine and are readily quantified by LC-MS/MS using standardized and validated methods.[9] Structures of NNK, NNAL, and NNN are shown in Fig. 4.3. There are several advantages of this biomarker: it is a metabolite of a powerful lung carcinogen, it is completely tobacco-specific, and it is present in virtually all tobacco products. Thus total NNAL in urine indicates the dose of a lung carcinogen. All smokers have total NNAL in their urine, and it has been quantified in a number of relatively large studies.[10–13] Another advantage is its relatively long lifetime in urine, facilitating identification of exposures to cigarette smoke that may have occurred more than 1 week earlier. Total NNAL has also been used widely to assess second-hand smoke exposure.[9,14,15] NNAL can also be quantified in plasma or toenails.[9] Total NNN in urine has not been as widely applied, as its concentrations are lower and there are some challenges in its measurement. Nevertheless, it provides a good biomarker of a tobacco-specific oral cavity and esophageal carcinogen, which is particularly important for studies of smokeless tobacco use.[9]

Another type of tobacco-specific biomarker is 4-hydroxy-1-(3-pyridyl)-1-butanone (HPB, Fig. 4.3) released from hemoglobin or DNA. Thus, NNK and NNN, upon metabolism, transfer the 4-oxo-4-(3-pyridyl) moiety to globin or DNA, and this can be released by hydrolysis and quantified by LC-MS/MS. Analysis of oral cell DNA from smokers and nonsmokers demonstrated significantly higher levels of released HPB in smokers.[16]

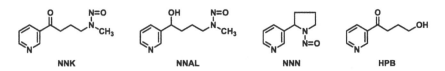

NNK   NNAL   NNN   HPB

Figure 4.3 Structures of 4-(methylnitrosamino)-1-(3-pyridyl)-1-butanone (NNK), 4-(methylnitrosamino)-1-(3-pyridyl)-1-butanol (NNAL), N'-nitrosonornicotine (NNN), and 4-hydroxy-1-(3-pyridyl)-1-butanone (HPB).

Volatile nitrosamines such as N-nitrosodimethylamine and N-nitrosopyrrolidine have been detected in smokers' urine, with levels reported to be higher in smokers than in nonsmokers, but the levels are low and relatively few studies have been performed.[17]

Polycyclic aromatic hydrocarbons (PAH) are formed during the incomplete combustion of organic matter. PAH always occur as mixtures, and some PAH such as benzo[a]pyrene are powerful carcinogens. Cigarette smoke, broiled foods, polluted air, and certain industrial settings are common sources of PAH exposure.[18] PAH are converted to epoxides and phenolic derivatives during metabolism frequently catalyzed by cytochromes P450 1A1, 1A2, and 1B1. Metabolites of PAH with 3 or 4 rings are excreted in the urine and are useful biomarkers of PAH exposure. Thus commonly measured urinary PAH biomarkers include 1-hydroxypyrene (1-HOP), 1-, 2-, 3-, 4-, and 9-hydroxyphenanthrene, 2-, 3-, and 9-hydroxyfluorene, and 1- and 2-hydroxynaphthalene (see Fig 4.4). Although not tobacco-specific, levels of all of these biomarkers are generally significantly elevated in smokers compared to nonsmokers. Thus, geometric means in ng/mL from the NHANES 2001−02 data were (smokers vs nonsmokers): 1-HOP (104 vs 40); 3-hydroxyphenanthrene (194 vs 91); 2-hydroxyfluorene (990 vs 236); and 1-hydroxynaphthalene (6293 vs 1523).[19,20] Another useful PAH biomarker is the urinary phenanthrene metabolite r-1,t-2,3,c-4-tetrahydroxy-1,2,3,4-tetrahydrophenanthrene (PheT, Fig. 4.5), which mimics the metabolic activation process of benzo[a]pyrene to its carcinogenic metabolite and is consequently a biomarker of PAH exposure plus metabolic activation.[10,21,22] Levels of

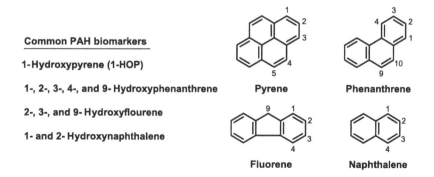

Common PAH biomarkers

1-Hydroxypyrene (1-HOP)

1-, 2-, 3-, 4-, and 9- Hydroxyphenanthrene

2-, 3-, and 9- Hydroxyflourene

1- and 2- Hydroxynaphthalene

Pyrene   Phenanthrene

Fluorene   Naphthalene

Figure 4.4 Commonly measured urinary polycyclic aromatic hydrocarbon (PAH) biomarkers. From Li Z, Sandau CD, Romanoff LC, Caudill SP, Sjodin A, Needham LL et al. Concentration and profile of 22 urinary polycyclic aromatic hydrocarbon metabolites in the U.S. population. *Environ Res* 2008;107(3):320−31.

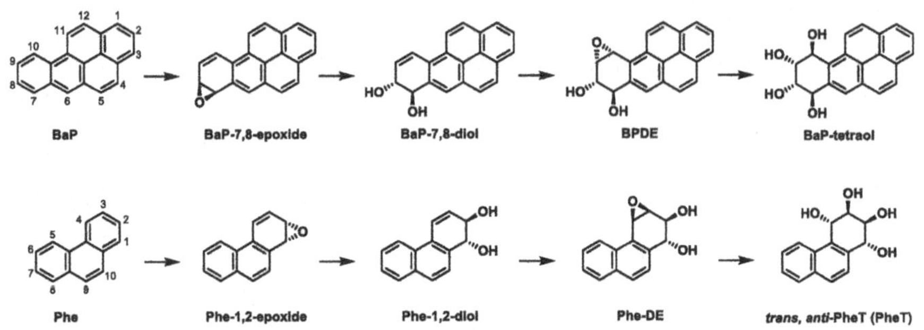

Figure 4.5 Metabolic activation of benzo[a]pyrene (BaP) and phenanthrene (Phe). From Hecht SS, Chen M, Yagi H, Jerina DM, Carmella SG. *r*-1,*t*-2,3,*c*-4-Tetrahydroxy-1,2,3,4-tetrahydrophenanthrene in human urine: a potential biomarker for assessing polycyclic aromatic hydrocarbon metabolic activation. *Cancer Epidemiol Biomarkers & Prev* 2003;**12**:1501−8.

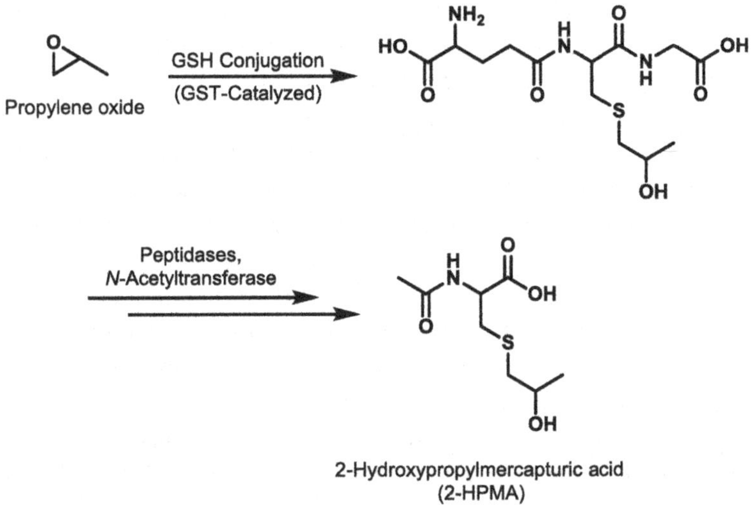

Figure 4.6 Formation of urinary mercapturic acids (propylene oxide as an example).

PheT in smokers in the United States were typically about three times higher than in nonsmokers and correlated with the corresponding tetraol resulting from the metabolic activation of benzo[a]pyrene.[21,23,24]

Cigarette smoking produces a variety of volatile toxicants and carcinogens. While some of these compounds can be quantified in blood, the most useful and widely applied method for assessing exposure to them is measurement of their mercapturic acid metabolites in urine. The formation of mercapturic acids from propylene oxide, as an example, is illustrated in Fig. 4.6. The electrophilic metabolite of a toxicant is frequently inactivated by reaction with glutathione, catalyzed by glutathione *S*-transferases. The resulting glutathione conjugate is then further processed enzymatically yielding an *N*-acetylcysteine conjugate that is excreted in the urine. These *N*-acetylcysteine conjugates are called mercapturic acids.[25] Multiple mercapturic acids have been quantified in the urine of smokers and those derived from acrolein, acrylamide, acrylonitrile, benzene, 1,3-butadiene, crotonaldehyde, dimethylformamide, propylene oxide, styrene, and xylene are significantly elevated compared to levels in nonsmokers, although it should be noted that, like the PAH metabolites, virtually all human urine samples contain these compounds.[3,26,27] The relationship of certain mercapturic acids to smoking has been clearly demonstrating by monitoring their concentrations after people stop smoking; levels of the mercapturic acids of acrolein, crotonaldehyde, ethylene oxide, benzene, and 1,3-butadiene decreased precipitously upon smoking cessation.[28] Consistent with this observation, data from the NHANES study demonstrated that levels of the major mercapturic acid derived from the tobacco smoke toxicant acrolein were about five times higher in smokers than in nonsmokers and were significantly predicted by serum cotinine levels.[29]

Hemoglobin adducts of volatile cigarette smoke constituents such as acrylonitrile, acrylamide, and ethylene oxide have also proven to be useful and robust biomarkers for quantifying exposure to these compounds. A distinct advantage of hemoglobin adducts is their ability to provide longer term integrative assessment of exposure than urinary metabolites because of the relatively long lifetime of the red blood cell. Thus hemoglobin adducts of these volatiles are consistently elevated in smokers.[3] DNA adducts of volatiles such as acrolein, formaldehyde, and acetaldehyde have also been quantified in leukocytes of smokers.[30–32]

Aromatic amines are constituents of cigarette smoke associated with bladder cancer etiology in smokers, and possibly with other cancers. Related compounds include the heterocyclic aromatic amines with diverse carcinogenic properties. Measurement of the parent compounds in urine is one approach to assess aromatic amine exposure in smokers. Two aromatic amines—4-aminobiphenyl and 2-naphthylamine—are recognized human bladder carcinogens. Some studies have detected relatively small amounts of these two aromatic amines and their metabolites in human urine.[33] Some of the heterocyclic aromatic

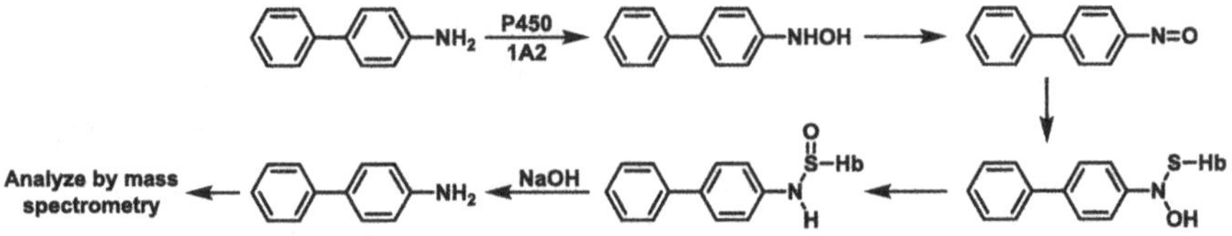

Figure 4.7 Formation and analysis of 4-aminobiphenyl-Hb adducts. From Tannenbaum SR, Skipper PL. Quantitative analysis of hemoglobin-xenobiotic adducts. *Methods Enzymol* 1994;**231**:625−32.

amines have also been quantified in human urine; levels of amino-α-carboline were consistently higher in the urine of smokers than nonsmokers.[33] Aromatic amine-hemoglobin adducts are the best established biomarkers of aromatic amine exposure in smokers. The formation of these adducts is outlined in Fig. 4.7. Validated methods for quantitation of these adducts have been used in multiple studies. The methods have been applied to a number of different aromatic amines including 3- and 4-aminobiphenyl, several dimethylanilines, as well as ethylanilines. Adduct levels are consistently higher in smokers than in nonsmokers.[33,34]

Tobacco smoke contains metals, some of which have powerful carcinogenic or toxic properties. NHANES data and other studies demonstrate that levels of the toxicant lead in blood and levels of the carcinogen cadmium in blood and urine are significantly higher in smokers than in nonsmokers.[35,36]

BIOMARKERS OF EXPOSURE IN E-CIGARETTE USERS

Exhaled CO and carboxyhemoglobin levels were significantly lower in e-Cigarette users than in cigarette smokers, based on the results of several studies.[37−40] This is consistent with the lack of combustion in e-Cigarette products compared to cigarette smoke. For example, in one study, exhaled CO levels in smokers were elevated above baseline and varied between 10 and 16 ppm up to 45 minutes after smoking a cigarette while no elevations in exhaled CO were observed in e-Cigarette users.[37] In another investigation, carboxyhemoglobin levels were significantly reduced after switching from tobacco cigarettes to e-Cigarettes for 2 weeks.[39]

Multiple studies have evaluated biomarkers of nicotine uptake in e-Cigarette users. Two reviews have summarized the results of studies published through January 2014.[41−43] These studies quantified plasma nicotine, plasma or serum cotinine, and salivary cotinine. Collectively, they demonstrate that e-Cigarette users are able to achieve concentrations of nicotine and cotinine similar to those of cigarette smokers, but this is dependent on the experience of the user and the product used. Experienced users could alter their use patterns to increase the amount of nicotine absorbed.[44] While the peak concentration of nicotine was similar to that of cigarette smokers, there may be a delay in the time taken to reach that peak.[41,42] One recent study demonstrated that in smokers with a week's worth of practice and 12 hours abstinence from nicotine, 10 puffs of an e-Cigarette over a 4.5-minute period caused increases in plasma nicotine and heart rate, and a decrease in craving, suggesting a clinically significant nicotine boost.[45] A longitudinal study examined salivary cotinine concentrations in former smokers who were using e-Cigarettes daily. Their cotinine levels were similar to those usually observed in cigarette smokers and they maintained these levels over an 8-month period by varying their consumption of e-liquid.[46] In a study of heavy smokers who abstained for 6 hours prior to e-Cigarette use, plasma nicotine and cotinine concentrations were measured at 10-minute intervals for 90 minutes while the subjects used a modified e-Cigarette. The results showed that e-Cigarettes deliver nicotine effectively, but with pharmacokinetic profiles lower than those obtained using tobacco cigarettes.[47] Urinary cotinine levels were generally similar in e-Cigarette users and smokers, based on the results of three recent studies.[38,40,48]

Levels of total NNAL were quantified in the urine of 28 e-Cigarette users who reported not smoking tobacco cigarettes for at least 2 months (confirmed by exhaled CO), using e-Cigarettes for at least 1 month and at least 4 days/week, with no current use of medicinal nicotine or other tobacco products.[40] The geometric mean of total NNAL was 0.02 pmol/mL urine, which was significantly lower than the geometric means of 1.48 and 1.21 pmol/mL urine in two studies of cigarette smokers (see Table 4.1). Total NNAL was below the limit of detection in 16 of the subjects. These results reflect the fact that the amounts of NNK reported in e-liquids are generally quite low,[51] because the main source of NNK in cigarette smoke is transfer from tobacco, where it is formed during curing and processing.[8] However, relatively high total NNAL values were obtained in the analysis of urine samples from four of the e-Cigarette users in this study, for reasons that are unclear and require further investigation.

Table 4.1 Geometric Means and 95% Confidence Intervals of Levels of Toxicant and Carcinogen Metabolites in the Urine of e-Cigarette Users and Cigarette Smokers by Study, Adjusted for Age and Gender

| Metabolite | e-Cigarettes | Cigarette Smokers | | | | | |
| | e-Cigarette Users N = 28 | Carmella et al. (2009)[28] N = 17 | | Hatsukami et al. (2010)[49] N = 165 | | Zarth et al. (2014)[50] N = 40 | |
	Geometric Mean (95% CI)	Geometric Mean (95% CI)	p-Value[a]	Geometric Mean (95% CI)	p-Value[a]	Geometric Mean (95% CI)	p-Value[a]
1-HOP (pmol/mL)	0.38 (0.26–0.55)	0.88 (0.55–1.41)	0.013	0.97 (0.80–1.17)	<0.0001	Not analyzed	
Total NNAL (pmol/mL)	0.02 (0.02–0.03)	1.48 (0.90–2.43)	<0.0001	1.21 (0.99–1.47)	<0.0001	Not analyzed	
3-HPMA (pmol/mL)	1100 (766–1590)	5800 (3730–9030)	<0.0001	4040 (3380–4830)	<0.0001	6070 (4580–8050)	<0.0001
2-HPMA (pmol/mL)	141 (80–252)	Not analyzed		Not analyzed		399 (255–626)	0.006
HMPMA (pmol/mL)	705 (456–1090)	4990 (2930–8490)	<0.0001	Not analyzed		Not analyzed	
SPMA (pmol/mL)	0.29 (0.18–0.46)	1.11 (0.61–2.08)	0.001	2.85 (2.24–3.63)	<0.0001	Not analyzed	
Nicotine (ng/mL)	869 (604–1250)	Not analyzed		1380 (1190–1600)	0.035	1270 (834–1710)	0.182
Cotinine (ng/mL)	1880 (1420–2480)	Not analyzed		3930 (3500–4400)	<0.0001	1930 (1530–2440)	0.981

[a]Compared to e-Cigarette users, adjusted for age and gender.

Source: From Hecht SS, Carmella SG, Kotandeniya D, Pillsbury ME, Chen M, Ransom BWS et al. Evaluation of toxicant and carcinogen metabolites in the urine of e-Cigarette users versus cigarette smokers. Nicotine Tob Res 2015;17:704–09, <http://dx.doi.org/10.1093/ntr/ntu218>.

Total NNN was also quantified in the urine samples analyzed for total NNAL. Similar to the data for total NNAL, levels of total NNN were 0.0055 pmol/mL, significantly lower than the average value of 0.06 pmol/mL urine found in cigarette smokers. These results also reflect the relatively low levels of NNN which have been reported in e-liquids.[51]

Levels of the PAH biomarker 1-HOP were significantly lower in the urine of 28 e-Cigarette users than in cigarette smokers (Table 4.1). Thus the geometric mean was 0.38 pmol/mL, significantly lower than the geometric means in two studies of cigarette smokers—0.88 and 0.97 pmol/mL.[40] Levels of 1-HOP in the NHANES study of nonsmokers were 0.18 pmol/mL. These results indicate that there is little formation of PAH during the use of e-Cigarettes, consistent with the presumed lack of combustion and the low levels of CO found in the exhaled breath of e-Cigarette users. The levels of 1-HOP which were detected are probably due to dietary and environmental exposure to PAH, which are ubiquitous.

Two studies have examined the effects of e-Cigarette use on levels of 3-hydroxypropyl mercapturic acid (3-HPMA), a metabolite of the powerful toxicant acrolein. In one, there was a statistically significant 79% reduction in levels of 3-HPMA in smokers who switched to e-Cigarette use for 4 weeks and remained abstinent from cigarette smoking. There was also a significant 60% reduction in dual users of cigarettes and e-Cigarettes.[38] A second study reported similar results; geometric mean levels of 3-HPMA were 1100 pmol/mL urine, significantly less than the values of 5800 pmol/mL urine and 4040 pmol/mL urine obtained in two studies of cigarette smokers.[40] Similarly lower levels of the mercapturic acids 2-HPMA (from propylene oxide), HMPMA (from crotonaldehyde), and SPMA (from benzene) were found in users of e-Cigarettes versus cigarette smokers (Table 4.1).[40]

There have been no reports in the literature on levels of biomarkers of aromatic amines or metals in e-Cigarette users.

SUMMARY

The results of biomarker studies reported to date indicate that the use of e-Cigarettes results in uptake of nicotine, generally at similar levels as in cigarette smokers, although the pharmacokinetics may differ.

Most results also indicate that this method of obtaining nicotine exposes the user to considerably lower levels of a variety of toxicants and carcinogens including carbon monoxide, tobacco-specific nitrosamines, PAH, and volatiles such as acrolein and benzene. These results are consistent with analyses of e-Cigarette liquids and vapor which have generally shown very low levels of many of the toxicants found in cigarette smoke. Nevertheless, caution is still advisable in the evaluation of toxicant exposure from e-Cigarettes, as the number and breadth of the published studies so far are rather limited. For example, two studies indicate that formaldehyde concentrations in e-Cigarette vapor may be elevated under certain conditions of use, but no biomarker studies of formaldehyde exposure in e-Cigarette users have been reported.[52,53] There have also been no reports of biomarkers of oxidative damage and inflammation in e-Cigarette users versus cigarette smokers. Thus further studies are warranted to gain a full appreciation of the relative merits of e-Cigarette use versus cigarette smoking with respect to exposure to toxicants and carcinogens.

REFERENCES

1. Walton KM, Abrams DB, Bailey WC, Clark D, Connolly GN, Djordjevic MV, et al. NIH electronic cigarette workshop: developing a research agenda. *Nicotine Tob Res* 2015; **17**(2):259−69.

2. Scherer G. Carboxyhemoglobin and thiocyanate as biomarkers of exposure to carbon monoxide and hydrogen cyanide in tobacco smoke. *Exp Toxicol Pathol* 2006;**58**(2−3):101−24.

3. Hecht SS, Yuan J-M, Hatsukami DK. Applying tobacco carcinogen and toxicant biomarkers in product regulation and cancer prevention. *Chem Res Toxicol* 2010;**23**:1001−8. Available from: http://dx.doi.org/10.1021/tx100056m.

4. Murphy SE, Park S-SL, Thompson EF, Wilkens LR, Patel Y, Stram DO, et al. Nicotine *N*-glucuronidation relative to *N*-oxidation and *C*-oxidation and UGT2B10 genotype in five ethnic/racial groups. *Carcinogenesis* 2014;**35**:2526−33.

5. von Weymarn LB, Thomson NM, Donny EC, Hatsukami DK, Murphy SE. Quantitation of the minor tobacco alkaloids nornicotine, anatabine, and anabasine in smokers' urine by high throughput liquid chromatography-mass spectrometry. *Chem Res Toxicol* 2016;**29**(3):390−7.

6. Hecht SS. Biochemistry, biology, and carcinogenicity of tobacco-specific *N*-nitrosamines. *Chem Res Toxicol* 1998;**11**:559−603.

7. Hecht SS. Progress and challenges in selected areas of tobacco carcinogenesis. *Chem Res Toxicol* 2008;**21**:160−71.

8. International Agency for Research on Cancer. *Smokeless tobacco and tobacco-specific nitrosamines. IARC Monographs on the Evaluation of Carcinogenic Risks to Humans*, vol. 89. Lyon, FR: IARC; 2007p. 421−583.

9. Hecht SS, Stepanov I, Carmella SG. Exposure and metabolic activation biomarkers of carcinogenic tobacco-specific nitrosamines. *Acc Chem Res* 2016;**49**(1):106−14.

10. Carmella SG, Ming X, Olvera N, Brookmeyer C, Yoder A, Hecht SS. High throughput liquid and gas chromatography-tandem mass spectrometry assays for tobacco-specific nitrosamine and polycyclic aromatic hydrocarbon metabolites associated with lung cancer in smokers. *Chem Res Toxicol* 2013;26(8):1209–17. Available from: http://dx.doi.org/10.1021/tx400121n.

11. Xia Y, Bernert JT, Jain RB, Ashley DL, Pirkle JL. Tobacco-specific nitrosamine 4-(methylnitrosamino)-1-(3-pyridyl)-1-butanol (NNAL) in smokers in the United States: NHANES 2007-2008. *Biomarkers* 2011;16(2):112–19.

12. Roethig HJ, Munjal S, Feng S, Liang Q, Sarkar M, Walk RA, et al. Population estimates for biomarkers of exposure to cigarette smoke in adult U.S. cigarette smokers. *Nicotine Tob Res* 2009;11(10):1216–25.

13. Vogel RI, Carmella SG, Stepanov I, Hatsukami DK, Hecht SS. The ratio of a urinary tobacco-specific lung carcinogen metabolite to cotinine is significantly higher in passive than in active smokers. *Biomarkers* 2011;16:491–7. Available from: http://dx.doi.org/10.3109/1354750X.2011.598565.

14. Wei B, Blount BC, Xia B, Wang L. Assessing exposure to tobacco-specific carcinogen NNK using its urinary metabolite NNAL measured in US population: 2011–2012. *J Expo Sci Environ Epidemiol* 2015.

15. Bernert JT, Pirkle JL, Xia Y, Jain RB, Ashley DL, Sampson EJ. Urine concentrations of a tobacco-specific nitrosamine carcinogen in the U.S. population from secondhand smoke exposure. *Cancer Epidemiol Biomarkers Prev* 2010;19(11):2969–77.

16. Stepanov I, Muzic J, Le CT, Sebero E, Villalta P, Ma B, et al. Analysis of 4-hydroxy-1-(3-pyridyl)-1-butanone (HPB)-releasing DNA adducts in human exfoliated oral mucosa cells by liquid chromatography-electrospray ionization-tandem mass spectrometry. *Chem Res Toxicol* 2013;26:37–45.

17. Seyler TH, Kim JG, Hodgson JA, Cowan EA, Blount BC, Wang L. Quantitation of urinary volatile nitrosamines from exposure to tobacco smoke. *J Anal Toxicol* 2013;37(4):195–202.

18. International Agency for Research on Cancer. *Some non-heterocyclic polycyclic aromatic hydrocarbons and some related exposures. IARC Monographs on the Evaluation of Carcinogenic Risks to Humans*, vol. 92. Lyon, FR: IARC; 2010p. 35–818.

19. Li Z, Sandau CD, Romanoff LC, Caudill SP, Sjodin A, Needham LL, et al. Concentration and profile of 22 urinary polycyclic aromatic hydrocarbon metabolites in the U.S. population. *Environ Res* 2008;107(3):320–31.

20. Suwan-ampai P, Navas-Acien A, Strickland PT, Agnew J. Involuntary tobacco smoke exposure and urinary levels of polycyclic aromatic hydrocarbons in the United States, 1999 to 2002. *Cancer Epidemiol Biomarkers Prev* 2009;18(3):884–93.

21. Hecht SS, Chen M, Yagi H, Jerina DM, Carmella SG. r-1,t-2,3,c-4-Tetrahydroxy-1,2,3,4-tetrahydrophenanthrene in human urine: a potential biomarker for assessing polycyclic aromatic hydrocarbon metabolic activation. *Cancer Epidemiol Biomarkers Prev* 2003;12:1501–8.

22. Zhong Y, Wang J, Carmella SG, Hochalter JB, Rauch D, Oliver A, et al. Metabolism of [D_{10}]phenanthrene to tetraols in smokers for potential lung cancer susceptibility assessment: Comparison of oral and inhalation routes of administration. *J Pharmacol Exp Ther* 2011;338:353–61.

23. Hecht SS, Carmella SG, Yoder A, Chen M, Li Z, Le C, et al. Comparison of polymorphisms in genes involved in polycyclic aromatic hydrocarbon metabolism with urinary phenanthrene metabolite ratios in smokers. *Cancer Epidemiol Biomarkers Prev* 2006;15:1805–11.

24. Hochalter JB, Zhong Y, Han S, Carmella SG, Hecht SS. Quantitation of a minor enantiomer of phenanthrene tetraol in human urine: correlations with levels of overall phenanthrene tetraol, benzo[a]pyrene tetraol, and 1-hydroxypyrene. *Chem Res Toxicol* 2011;24(2):262–8. Available from: http://dx.doi.org/10.1021/tx100391z.

25. Mathias PI, B'Hymer C. A survey of liquid chromatographic-mass spectrometric analysis of mercapturic acid biomarkers in occupational and environmental exposure monitoring. *J Chromatogr B Analyt Technol Biomed Life Sci* 2014;**964**:136−45.

26. Pluym N, Gilch G, Scherer G, Scherer M. Analysis of 18 urinary mercapturic acids by two high-throughput multiplex-LC-MS/MS methods. *Anal Bioanal Chem* 2015;**407**(18):5463−76.

27. Alwis KU, Blount BC, Britt AS, Patel D, Ashley DL. Simultaneous analysis of 28 urinary VOC metabolites using ultra high performance liquid chromatography coupled with electrospray ionization tandem mass spectrometry (UPLC-ESI/MSMS). *Anal Chim Acta* 2012; **750**:152−60.

28. Carmella SG, Chen M, Han S, Briggs A, Jensen J, Hatsukami DK, et al. Effects of smoking cessation on eight urinary tobacco carcinogen and toxicant biomarkers. *Chem Res Toxicol* 2009;**22**(4):734−41. Available from: http://dx.doi.org/10.1021/tx800479s.

29. Alwis KU, deCastro BR, Morrow JC, Blount BC. Acrolein exposure in U.S. tobacco smokers and non-tobacco users: NHANES 2005-2006. *Environ Health Perspect* 2015; **123**(12):1302−8.

30. Wang M, Cheng G, Balbo S, Carmella SG, Villalta PW, Hecht SS. Clear differences in levels of a formaldehyde-DNA adduct in leukocytes of smokers and non-smokers. *Cancer Res* 2009;**69**:7170−4.

31. Chen L, Wang M, Villalta PW, Luo X, Feuer R, Jensen J, et al. Quantitation of an acetaldehyde adduct in human leukocyte DNA and the effect of smoking cessation. *Chem Res Toxicol* 2007;**20**:108−13.

32. Zhang S, Balbo S, Wang M, Hecht SS. Analysis of acrolein-derived 1,N^2-propanodeoxyguanosine adducts in human leukocyte DNA from smokers and nonsmokers. *Chem Res Toxicol* 2011;**24**(1):119−24.

33. Turesky RJ, Le Marchand L. Metabolism and biomarkers of heterocyclic aromatic amines in molecular epidemiology studies: lessons learned from aromatic amines. *Chem Res Toxicol* 2011;**24**(8):1169−214.

34. Gan J, Skipper PL, Gago-Dominguez M, Arakawa K, Ross RK, Yu MC, et al. Alkylaniline-hemoglobin adducts and risk of non-smoking-related bladder cancer. *J Natl Cancer Inst* 2004;**96**(19):1425−31.

35. Richter PA, Bishop EE, Wang J, Kaufmann R. Trends in tobacco smoke exposure and blood lead levels among youths and adults in the United States: the National Health and Nutrition Examination Survey, 1999−2008. *Prev Chronic Dis.* 2013;**10**:E213.

36. Jones MR, Apelberg BJ, Tellez-Plaza M, Samet JM, Navas-Acien A. Menthol cigarettes, race/ethnicity, and biomarkers of tobacco use in U.S. adults: the 1999−2010 National Health and Nutrition Examination Survey (NHANES). *Cancer Epidemiol Biomarkers Prev* 2013;**22**(2):224−32.

37. Vansickel AR, Cobb CO, Weaver MF, Eissenberg TE. A clinical laboratory model for evaluating the acute effects of electronic "cigarettes": nicotine delivery profile and cardiovascular and subjective effects. *Cancer Epidemiol Biomarkers Prev* 2010;**19**(8):1945−53.

38. McRobbie H, Phillips A, Goniewicz ML, Smith KM, Knight-West O, Przulj D, et al. Effects of switching to electronic cigarettes with and without concurrent smoking on exposure to nicotine, carbon monoxide, and acrolein. *Cancer Prev Res (Phila)* 2015;**8**(9):873−8.

39. van Staden SR, Groenewald M, Engelbrecht R, Becker PJ, Hazelhurst LT. Carboxyhaemoglobin levels, health and lifestyle perceptions in smokers converting from tobacco cigarettes to electronic cigarettes. *S Afr Med J* 2013;**103**(11):865−8.

40. Hecht SS, Carmella SG, Kotandeniya D, Pillsbury ME, Chen M, Ransom BWS, et al. Evaluation of toxicant and carcinogen metabolites in the urine of e-cigarette users versus cigarette smokers. *Nicotine Tob Res* 2015;**17**:704−9. Available from: http://dx.doi.org/10.1093/ntr/ntu218 Epub 2014 Oct 21. [Epub ahead of print].

41. Schroeder MJ, Hoffman AC. Electronic cigarettes and nicotine clinical pharmacology. *Tob Control* 2014;**23**(Suppl 2):ii30−5.

42. Marsot A, Simon N. Nicotine and cotinine levels with electronic cigarette: A review. *Int J Toxicol* 2016;**35**(2):179−85.

43. Grana R, Benowitz N, Glantz SA. E-cigarettes: a scientific review. *Circulation* 2014; **129**(19):1972−86.

44. Farsalinos KE, Romagna G, Tsiapras D, Kyrzopoulos S, Voudris V. Evaluation of electronic cigarette use (vaping) topography and estimation of liquid consumption: implications for research protocol standards definition and for public health authorities' regulation. *Int J Environ Res Public Health* 2013;**10**(6):2500−14.

45. Nides MA, Leischow SJ, Bhatter M, Simmons M. Nicotine blood levels and short-term smoking reduction with an electronic nicotine delivery system. *Am J Health Behav* 2014; **38**(2):265−74.

46. Etter JF. A longitudinal study of cotinine in long-term daily users of e-cigarettes. *Drug Alcohol Depend* 2016;**160**:218−21.

47. Velez dM, Jones DR, Jahn A, Bies RR, Brown JW. Nicotine and cotinine exposure from electronic cigarettes: a population approach. *Clin Pharmacokinet* 2015;**54**(6):615−26.

48. Goney G, Cok I, Tamer U, Burgaz S, Sengezer T. Urinary cotinine levels of electronic cigarette (e-cigarette) users. *Toxicol Mech Methods.* 2016;1−5.

49. Hatsukami DK, Kotlyar M, Hertsgaard LA, Zhang Y, Carmella SG, Jensen JA, et al. Reduced nicotine content cigarettes: effects on toxicant exposure, dependence and cessation. *Addiction* 2010;**105**:343−55. Available from: http://dx.doi.org/10.1111/j.1360-0443.2009.02780.x.

50. Zarth A, Carmella SG, Le CT, Hecht SS. Effect of cigarette smoking on urinary 2-hydroxypropylmercapturic acid, a metabolite of propylene oxide. *J Chromatog B* 2014;**953−954**:126−31. Available from: http://dx.doi.org/10.1016/j.jchromb.2014.02.001.

51. Famele M, Ferranti C, Abenavoli C, Palleschi L, Mancinelli R, Draisci R. The chemical components of electronic cigarette cartridges and refill fluids: review of analytical methods. *Nicotine Tob Res* 2015;**17**(3):271−9.

52. Kosmider L, Sobczak A, Fik M, Knysak J, Zaciera M, Kurek J, et al. Carbonyl compounds in electronic cigarette vapors-effects of nicotine solvent and battery output voltage. *Nicotine Tob Res* 2014;**16**:1319−26. Available from: http://dx.doi.org/10.1093/ntr/ntu078.

53. Jensen RP, Luo W, Pankow JF, Strongin RM, Peyton DH. Hidden formaldehyde in e-cigarette aerosols. *N Engl J Med* 2015;**372**(4):392−4.

CHAPTER 5

Review of Compounds of Regulatory Concern

K.E. Farsalinos

INTRODUCTION

As with any consumer product, regulation for e-Cigarettes is important in order to ensure the best possible quality, safety, consistency in product standards and contents, and effectiveness for the intended use. However, the issue of e-Cigarette regulation is more complex than for other consumer products. e-Cigarette use is associated with the intake of an addictive substance, nicotine. Also, the pattern of use closely resembles the act of smoking, and the intended use from a public health perspective is as a smoking substitute and not as a new "trend" or habit. As a result, the regulatory assessment of e-Cigarettes is much more complex and challenging in terms of their content and emissions as well as for promotion, marketing, and related public perception of the product. It is tempting to regulate e-Cigarettes based on absolute safety and using the precautionary principle. That would minimize the risk of avoidable harm, related to exposure to toxins in e-Cigarette, renormalization, gateway progression to smoking, or other real or potential risks.[1] However, if this approach also makes e-Cigarettes less easily accessible, less palatable or acceptable, more expensive, less consumer friendly or pharmacologically less effective for smokers, or if it inhibits innovation and development of new and improved products, then it causes harm by perpetuating smoking.[1] Getting this balance right will be a challenging process.

This chapter focuses on the regulatory issues related to the chemical composition of e-Cigarette liquids and aerosol, and will review the current evidence on the main chemicals that could be targeted for regulation and control.

Analytical Assessment of e-Cigarettes. DOI: http://dx.doi.org/10.1016/B978-0-12-811241-0.00005-X

TOBACCO CIGARETTE REGULATION

The main focus of the regulation of tobacco cigarettes is to minimize appeal, reduce addictiveness, ban flavors, ban active advertising and promotion, and apply packaging and labeling rules that will properly inform the public about the health impact of smoking. Little of that is achieved by regulating the chemistry of the product or the smoke. Moreover, there is limited potential to regulate the main ingredient of the tobacco cigarette, which is the tobacco plant, besides applying techniques such as genetic engineering. Limits have been set for nicotine, tar, and carbon monoxide emissions in the smoke in the European Union (EU) and several other countries.[2] However, there is a lot of room to regulate the numerous additives that are introduced into tobacco cigarettes.

Tobacco cigarette smoke is a toxic product of combustion. It is a very complex and dynamic mixture, containing more than 5000 chemicals.[3,4] Additives were hardly used before 1970, but now they represent up to 10% of the cigarette weight.[5] It has been reported that almost 600 chemicals are used as additives to improve taste and reduce harshness of the smoke.[6] They are mainly sugars, humectants, cocoa, and liquorice.[5] Certain characteristic flavors, such as candy and fruit, have also been used and have made tobacco products more appealing to children.[7] To reduce that, the US Food and Drug Administration (FDA) banned several flavors as additives in September 2009,[8] and the EU has banned additives that result in a product with a characteristic flavor.[9] For all additives, the regulatory interest is mainly on the potentiation of addictiveness and appeal rather than on safety concerns. For added humectants, mainly propylene glycol and glycerol, research has shown that they do not increase the toxicity of the smoke.[10-12] Several other additives have been shown not to alter the toxicological profile of smoke.[13-16] However, sugars (naturally present in tobacco but also used as additives to reduce harshness) can alter the smoke chemistry, resulting in elevated emissions of toxic aldehydes with potential implications in toxicity and addictiveness.[17-22]

Efforts to regulate tobacco chemistry are only just emerging. The World Health Organisation Framework Convention on Tobacco Control (WHO FCTC) recognizes the need for a strategy to regulate tobacco products to reduce their attractiveness but does not yet provide any guidance for reducing either the dependence potential or toxicity of cigarettes and their ingredients.[23] In 2007, the WHO Study Group on Tobacco Product Regulation (TobReg) released a proposal, developed by a joint International Agency for Research on Cancer (IARC) and WHO working group, which presented performance standards for cigarettes and a strategy to use these standards to mandate a reduction in the toxicant yields of smoke.[24] This new approach was deemed necessary because of scientific evidence that the most common measurements to categorize cigarette smoke, machine-measured tar, nicotine and carbon monoxide yields per cigarette based on the US Federal Trade Commission (FTC)/International Standards Organization (ISO) testing regimen, do not provide valid estimates of human exposure or of relative human exposure for different brands of cigarettes.[24–26] The goal of TobReg was to reduce toxicant levels in mainstream smoke measured under standardized conditions and recommended establishing levels for selected toxicants per mg nicotine and prohibiting the sale or import of brands that have yields above these levels. A secondary goal was to prevent the introduction into a market of new tobacco cigarettes with higher levels of smoke toxicants than are present in brands already on the market. They released a priority list of nine toxicants, including tobacco-specific nitrosamines, aldehydes, and polycyclic aromatic hydrocarbons, and recommended specific limits based on analytical measurements of market products. The whole strategy was based on the implementation of a policy change in phases, beginning with a period of required annual reporting of toxicant levels by cigarette manufacturers to the regulatory authorities. This would be followed by the promulgation of the levels for toxicants above which brands cannot be offered for sale. The established levels would be enforced. In the United States, FDA published a list of harmful and potentially harmful constituents (HPHCs) in tobacco products, with a requirement that all manufacturers and importers report the amount of these chemicals in their cigarettes or smoke. The full list comprises 93 chemicals, but FDA issued draft guidance in March 2012 that identified a subset of 20 HPHCs for which manufacturers and importers are to test and report to FDA.[27] The main goal is to help people understand the potential harms of tobacco use, but the scheme can also be used to prevent the introduction of new products into the market by using the Substantial Equivalence regulation.[28] Another ongoing effort in the United States is to evaluate possibilities to reduce nicotine content of cigarettes to nondependence levels,[29–33] but no relevant regulatory decisions have been made. Recently, the WHO has also recommended the implementation of product regulations requiring reduction of nicotine levels.[34] In 2014, the EU set up a new Tobacco Product Directive (TPD) to regulate tobacco products.[9] One of the aspects by which the EU aims to regulate these products is by influencing the dependence potential. The TPD prohibits tobacco products with increased dependence potential, thereby specifically focusing on the role of additives or a combination of additives in increasing the dependence potential of cigarettes and roll-your-own tobacco. In 2010, the EU Scientific Committee on Emerging and Newly Identified Health Risks (SCENIHR) released a report evaluating the role of additives in the addictiveness and attractiveness of tobacco products.[5] A number of compounds of concern were analyzed, including sugars (which are pyrolyzed to aldehydes), ammonia and other pH modifiers, and flavoring compounds. The report concluded that it remains difficult to distinguish the direct effects of these additives from indirect effects such as marketing towards specific groups.

In conclusion, several attempts have been made to regulate tobacco cigarettes but almost all are related to reducing appeal or addictiveness. The efforts to reduce toxin exposure through regulatory limitation of specific chemicals of concern are still evolving but are bound to be limited by the combustion process, which is unavoidable. This makes the creation of a "safe" tobacco cigarette a very difficult, if not impossible, task.

e-CIGARETTE REGULATION

Unlike tobacco cigarettes, there are many possibilities to regulate e-Cigarettes in an attempt to make them as safe as possible for the consumer. There are three main areas where regulation of the chemistry of the e-Cigarettes can have an impact: composition of the liquids, aerosol emissions, and addictive potential.

COMPOSITION OF e-CIGARETTE LIQUIDS

No chemical specifically developed de novo for use in e-Cigarette liquids has been reported. The main ingredients, besides nicotine, in most if not all liquids are propylene glycol, glycerol, and flavorings. It is reasonable for regulatory efforts to focus initially on the main ingredients.

Propylene Glycol

Propylene glycol ($C_3H_8O_2$, CAS Reg. No. 57-55-6) is also known as 1,2-propanediol (Fig. 5.1A). It is a synthetic compound that does not occur in nature, a viscous, colorless, and odorless liquid with a slightly sweet taste. Chemically it is classed as a diol (a polyhydric alcohol) due to the presence of two hydroxyl groups (OH). It has two enantiomers, levo-rotatory (L +) and dextro-rotatory (D −).[35] It is hygroscopic (having the ability to attract water molecules from the surrounding environment) and is thus classified as a humectant. It is totally miscible with water as well as with other solvents such as acetone and chloroform. Its boiling point is 184−186°C and its freezing point −60°C.[36]

Commercially, propylene glycol is mainly produced by hydrolysis of propylene oxide. It was first produced in 1859 by Charles-Adolphe Wurtz, one of the most important chemists of the 19th century.[37] It was commercialized in 1931 by Carbide and Carbon Chemicals Corporation. It has a significant role in the chemical industry and a wide range of applications. The production capacity in the United States was estimated at 1312 million pounds in 1998.[38] The most common use is in the production of unsaturated polyester resins,[39] but it is also used in products of human consumption. Uses and percent of

(A)

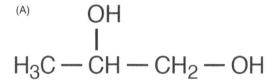

Propylene glycol (1,2-propanediol)

(B)

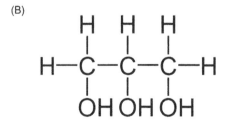

Glycerol(1,2,3-propanetriol)

Figure 5.1 Chemical structures of propylene glycol (A) and glycerol (B).

demand are as follows: unsaturated polyester resins, 26%; antifreeze and deicing fluids, 22%; food, drugs, and cosmetics, 18%; liquid detergents, 11%; functional fluids (inks, specialty antifreeze, deicing lubricants), 4%; pet foods, 3%; paints and coatings, 5%; tobacco, 3%; miscellaneous, including plasticizers, 8%.[38] It is also used in the generation of artificial mists and fogs used in fire safety training and theatrical productions.[40] A variety of intravenously and orally administered medications use it as a solvent, such as lorazepam, etomidate, phenytoin, diazepam, digoxin, hydralazine, esmolol, chlordiazepoxide, multivitamins, nitroglycerin, pentobarbital sodium, phenobarbital sodium, and trimethoprim-sulfamethoxazole.[41] It can also be added in respiratory inhalants to reduce the viscosity of bronchial secretions[42] and has been used as a vehicle for inhalation of immunosuppressive medications.[43,44] It is approved by the FDA for use in food, tobacco, and pharmaceutical products as an inert ingredient and is Generally Recognized As Safe (GRAS) for direct addition to foods.[45] It is used as an anticaking agent, antioxidant, dough strengthener, emulsifier, flavor agent, formulation aid, humectant, processing aid, solvent vehicle stabilizer and thickener, surface-active agent, and texturizer. It can be present at levels of 2–97%.[45] It should be noted that propylene glycol is available as both industrial and pharmaceutical (United States Pharmacopeia and European Pharmacopeia) grades.[39] There are specific standards for pharmaceutical grade: it should be of at least 99.5% purity, contain ≤5 ppm heavy metals, have a specific gravity of 1.035–1.040, and contain ≤0.2% water. Diethylene glycol and ethylene glycol impurities are not allowed at levels >0.10%.

The most common type of exposure to propylene glycol for the general population is through dermal and oral contact. Dermal contact can occur with consumer products such as cosmetics, antifreeze, coolants, windshield deicers and pharmaceutical creams. Oral exposure also occurs through use in food and tobacco products and as a solvent for pharmaceuticals.[46] In 2002, it was estimated that the per capita consumption of propylene glycol in the United States was 34.3 mg/kg body weight per day (approximately 2.4 g).[47] The average daily dietary intake in Japan was estimated to be 43 mg per person per day in 1982.[48] The WHO food additive series report in 1974 lists the acceptable human daily intake at <25 mg/kg body weight per day.[49] Other less common exposure pathways are from intravenous administration of pharmaceuticals and occupational exposure.[41] The latter may occur through direct dermal contact while handling products containing the compound or through inhalation of airborne propylene glycol resulting from heating or spraying processes.[41,46] Neither the US Occupational Safety and Health Administration (OSHA) nor the American Conference of Governmental Industrial Hygienists (ACGIH) has established exposure limits for propylene glycol vapor. There is also no Threshold Limit Value (TLV). The American Industrial Hygiene Association (AIHA) Workplace Environmental Exposure Level (WEEL) has determined an inhalation aerosol exposure safety limit of 10 mg/m^3.[50]

The main metabolic pathway in mammals is oxidation by alcohol dehydrogenase to lactaldehyde and then by aldehyde dehydrogenase to lactate.[51–54] Lactate is further metabolized to pyruvate and glucose (Fig. 5.2). An alternative possible metabolic pathway is phosphorylation to acetol phosphate, lactaldehyde phosphate, lactyl phosphate, and lactic acid.[55] In most mammals, part of the absorbed propylene glycol is eliminated unchanged by the kidneys, while another portion is excreted by the kidneys as a glucuronic acid conjugate.[56] About 45% of the absorbed propylene glycol is eliminated by the kidneys in humans.[57] The mean serum half-life is approximately 2–4 hours.[58]

Propylene glycol has very low systemic toxicity in experimental animals due to its metabolism to lactate and pyruvate, which are normal constituents of the citric acid cycle. Repeated exposure of rats to propylene glycol in drinking water or feed did not result in adverse effects at levels up to 10% in water (estimated at about 10 g/kg body weight per day) or 5% in feed (dosage reported as 2.5 g/kg per day) for periods up to 2 years.[46] Acute oral toxicity has been well characterized in the rat, mouse, rabbit, dog, and guinea pig, with LD$_{50}$ values of 8–46 g/kg. No fetal or developmental toxicity was observed in rats, mice, rabbits, or hamsters, with No Observed Adverse Effects Levels (NOAELs) ranging from 1.2 to 1.6 g/kg per day.[38] No lethal oral dose has been defined in humans, but it is estimated to be >15 g/kg.[46] No

Figure 5.2 Metabolism of propylene glycol in mammalian cells.

major differences in toxicity have been observed between humans and animals, with the only exception being cats, which developed Heinz body anemia in response to propylene glycol as an additive (at 6%, w/w, or above) to their diet.[41] There is no evidence of any carcinogenic effects in humans. It is mildly irritating to the eyes, and there are reports of positive skin reactions from exposure to propylene glycol patches.[59] The latter was mostly related to primary irritation rather than an allergic reaction.[41] In adults, toxicity has been observed at serum levels of >180 mg/L.[57] Central nervous system depression and seizures have been reported after repeated exposure at high levels, mostly from iatrogenic exposure and especially in children.[60–62] There is also the risk of lactic acidosis, usually from intravenous administration of medications which contain propylene glycol as solvent.[63–66]

In e-Cigarettes, it is widely used because of its excellent solvent properties and its production of a visible aerosol when heated. It was the solvent mentioned in the patent application of the inventor of e-Cigarettes.[67] It is the most important solvent used in the flavor concentrate industry; flavor concentrates are also used in e-Cigarette liquids. It is also responsible for throat irritation. In the case of smoking and e-Cigarette use, this is commonly called "throat hit," a desired effect for both smokers and e-Cigarette users;[68,69] a recent study showed that variables assessing satisfaction with e-Cigarettes were significantly associated with a stronger throat hit, indicating that consumers want to experience this effect.[70] Propylene glycol produces less visible aerosol but stronger throat hit and better flavor than other solvents used in e-Cigarette liquids.[71] However, excessive, throat hit can become an irritation even for e-Cigarette users, and this is reported as one of the most common side effects.[72]

e-Cigarettes have introduced a new pattern of exposure to propylene glycol for consumers: inhalation of aerosol produced by heating. Before that, inhalation exposure was only observed in deicing facilities in airports and in theatrical fog settings. The former was related to mist exposure, while the latter was associated with heating. As a result, research on inhalation is scarce. However, early studies on vapors were done in the 1940s, when it was found that the vapor had bactericidal properties and could protect animals from infection with the influenza virus.[73–75] Subsequent studies found a protective effect against common cold in humans.[76,77] The most extensive work on this issue was done by Robertson et al. from the University of Chicago.[78–80] In an attempt to evaluate the safety of inhaling propylene glycol vapors, this research group studied the effects of inhalation exposure on rats and monkeys. Continuous exposure to an atmosphere saturated with the vapor for 12–18 months showed no deleterious effect on any organ, including the lungs, compared to control animals.[81] A more recent study evaluated local respiratory and systemic effects in rats.[82] Researchers found elevation in the number of goblet cells and elevated mucin production by pre-existing goblet cells in the nasal turbinates, which was probably related to the dehydrating effects of propylene glycol and was not considered as an indication of irreversible damage. They concluded that no toxic effect was shown in any organ system or blood component. In humans, inhalation data are lacking. An experimental study by Weislander et al. evaluated the effects of exposure to the propylene glycol mist used for aircraft deicing in 27 nonasthmatic volunteers, 22% of whom were smokers and 44% former smokers.[83] The aim was to study effects of an experimental exposure at levels occurring during normal aviation emergency training. The effects studied included tear film stability, nasal patency, and lung function, as well as subjective symptoms. Lung function was evaluated by dynamic spirometry, measuring vital capacity (VC), forced vital capacity (FVC), peak expiratory flow (PEF), and forced expiratory flow in 1 second (FEV_1). The subjects were exposed to mist for 1 minute, with the geometric mean concentration of propylene glycol in the flight simulator being 309 mg/m^3. The most common symptoms, evaluated by questionnaire, were a sensation of sore and dry eyes, throat dryness, and irritative cough. A significant decrease of tear film stability was found after exposure, with a reduction of mean tear film stability breakup time from 38 to 29 seconds. This is probably related to the humectant effects of propylene glycol. Most of the lung function values remained unchanged after exposure, but there was a minor numerical decrease of FEV_1 from 103% at baseline to 102% after exposure, and a small but significant decrease of FEV_1/FVC ($P = 0.049$). Mean VC was unchanged, but FVC was slightly increased. None of the 27 participants had an initial lung function value (FEV_1) below 80% of predicted value, but one got a 77% value for FEV_1 after exposure. It should be noted that propylene glycol used in deicing facilities is of industrial grade[84] and thus of lower purity than that used in food products and pharmaceuticals. Moreover, the slight but statistically significant reduction in FEV_1/FVC was partly driven by an elevation in FVC. There is no disease condition which results in an elevation of FVC. In obstructive lung disorders, FEV_1 is disproportionally reduced compared to FVC, but they are both reduced.[85] In restrictive lung disorders, FVC is decreased at the same rate as FEV_1.[85] Therefore, the observation of reduced FEV_1/FVC does not appear to have any clinical significance.

Current evidence is limited but does not raise concerns about serious adverse effects from the use of propylene glycol in e-Cigarettes. It is not expected to be of concern when absorbed systemically, because of the well-known metabolic pathways and the relatively low amounts consumed daily (a few grams, depending on the liquid composition and daily consumption, much of which is exhaled). From a regulatory perspective, it is important to implement the use of pharmaceutical grade propylene glycol in e-Cigarette liquids, so that the levels of contaminants such as ethylene and diethylene glycol will be minimized. Furthermore, research on its safety for long-term inhalation should expand, while long-term epidemiological studies on e-Cigarette users will eventually evaluate its safety profile. The main focus of the research should be on the local effects on the respiratory tract. Finally, the use of alternative compounds with similar properties (acting as a solvent and producing visible aerosol) should be evaluated.

Glycerol

Glycerol ($C_3H_8O_3$, CAS Reg. No. 56-81-5) is also known as 1,2,3-propanetriol (Fig. 5.1B). It is a natural compound, a viscous, syrupy, colorless, and odorless liquid with a sweet taste. Chemically it is classed as a polyol (a polyhydric alcohol), due to the presence of three hydroxyl groups (OH). It is hygroscopic and thus a humectant (similarly to propylene glycol). It is miscible with water and alcohol and has a boiling point of 290°C and a freezing point of 17°C.[86]

Glycerol is essential for living organisms. It forms the backbone of triglycerides, esters derived from glycerol and three fatty acids,[87] which are present in humans, animals, and plants. It is the oldest organic molecule isolated by man, being obtained by heating fats in the presence of ash to produce soap since as early as 2800 BC.[88] As a substance, it was accidentally discovered in 1779 by K.W. Scheele, a Swedish chemist, while he was heating a mixture of olive oil and lead monoxide.[89] Scheele later established that other metals and glycerides produce the same chemical reaction that yields glycerol and soap, and in 1783 he published a description of his method of preparation in the transactions of the Royal Academy of Sweden. Scheele's method was used to produce glycerol commercially for some years. The name was derived from the Greek word *glykys*, meaning sweet, and was given by a French scientist, M.E. Chevreul, in the 19th century. In 1823, Chevreul obtained the first patent for a new way to produce fatty acids from fats treated with an alkali, which included the recovery of glycerol released during the process. Glycerol became a compound of strategic value when it was used in the manufacturing of nitroglycerine. In 1866, Alfred Nobel discovered that mixing nitroglycerin with the siliceous sedimentary rock kieselguhr, a form of silica mined in Germany, could turn the liquid into a malleable paste known as dynamite which could be kneaded and shaped into rods for blasting rock through a detonator activated by means of a strong shock. Glycerol shortly became a strategic military resource. Hence, when glycerol demand in the First World War exceeded the supply from the soap industry, reasons of military security led to the first synthetic plants to manufacture it in Europe and the United States, where glycerol for weaponry was produced through microbial sugar fermentation.[90]

Glycerol is made by saponification (hydrolysis) of animal fats and transesterification of vegetable fats (Fig. 5.3). Both processes yield 10% glycerol by weight as a byproduct. It can also be made synthetically from propylene using the epichlorohydrin process,[91] but this is not cost-effective. In 2000, the estimated world production of glycerol was 500,000 tons. In 1999, the amount imported and/or produced was 227,000 tons in Europe and around 28,000 tons in the

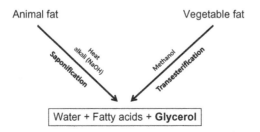

Figure 5.3 Glycerol production by saponification and transesterification of fats.

United Kingdom.[92] Production in the United States was estimated at 350,000 tons annually between 2000 and 2004.[93] More recent data suggest a worldwide annual production of 2 million tons, mostly as a result of the booming biodiesel and oleochemicals manufacturing.[94] As a result, supply has exceeded demand.

Glycerol has many applications. It is a constituent in numerous products and an intermediate in industrial applications for the manufacture of such things as soaps and detergents and glycerol esters. It is found in consumer items such as pharmaceuticals, cosmetics, tobacco, food and drinks, and in paints, resins, and paper. The most common uses are as intermediates and monomers in resins, polyols and polyurethanes (20%), in cosmetics (20%), as a chemical intermediate in the synthesis of other chemicals (15%), and in pharmaceuticals, tobacco, industrial fluids, cellulose films, and food products (<10% each).[92] It was first accorded GRAS status by the FDA in 1959.[89] It is currently listed as a multiple purpose GRAS food substance (21CFR 182.1320) and as a substance migrating from paper and paperboard products (21CFR 182.90) for use in certain food packaging materials. It is a humectant, solvent, sweetener, and preservative. Another important but indirect use in food processing is represented by monoglycerides, the glycerol esters of fatty acids, which are emulsifiers and stabilizers for many products.[89] There are specific standards for pharmaceutical grade, with small differences between United States and European Pharmacopeia definitions. It should be of at least 98.0% purity and contain ≤5 ppm heavy metals, ≤10 ppm chlorides, ≤30−35 ppm halogenated compounds, ≤2.0−5.0% water, and ≤ 10 ppm aldehydes.[95]

Data from human and animal studies indicate that glycerol is rapidly absorbed in the intestine and the stomach, distributed over the extracellular space, and excreted.[96] It is phosphorylated to alpha-glycerophosphate by glycerol kinase, predominantly in the liver (80−90%) and kidneys (10−20%), and incorporated in the standard metabolic pathways to form glucose and glycogen.[96,97] Glycerol kinase is also found in intestinal mucosa, brown adipose tissue, lymphatic tissue, lung, and pancreas. Glycerol may also be combined with free fatty acids in the liver to form triglycerides (lipogenesis) which are distributed to the adipose tissues. The turnover rate is directly proportional to plasma glycerol levels.[98]

Glycerol has been safely used in many industrial and pharmaceutical applications for over 100 years and is generally recognized for its low risk to health. Occupational exposure can occur during production, processing, or use of products containing it. The dermal route is considered to be the most relevant for exposure, although inhalation of aerosols released from the spray application of resins or paints may also occur.[92] The OSHA Permissible Exposure Limit (PEL) for glycerol mist inhalation is 15 mg/m^3 as total dust and 5 mg/m^3 for the respirable fraction (<10 μm particle size).[99] ACGIH has established a TLV of 10 mg/m^3. Consumer exposure to glycerol occurs principally through its use in food, cosmetics, toiletries, and pharmaceuticals, mainly dermally but also orally as a direct food additive, indirectly from cellulose films used for food applications, and via orally administered drugs and oral hygiene products. Additionally, there is limited consumer dermal exposure through contact with paints, printing inks, resins, and matrices containing glycerol.[92] Healthy individuals can easily tolerate oral doses of up to 1.5 g/kg or less with only slight diuresis occurring. Animal studies have reported oral LD$_{50}$ values of >24 g/kg in rats.[100,101] For mice and guinea pigs, the oral LD$_{50}$ values were 23 and 10 g/kg, respectively.[100] A study of repeated oral administration in rats for 2 years established a NOAEL level of 10 g/kg per day.[100] For acute dermal toxicity, a single LD$_{50}$ of >18 g/kg for rabbits has been established.[100] Glycerol is not considered a skin or eye irritant.[92] A study of intravenous administration in humans (to control cerebral edema) did not identify any toxic effects,[102] while a study of chronic ingestion found increased levels of plasma triglycerides.[103]

In e-Cigarettes, glycerol is mainly used for its solvent properties and because it produces a thick, highly visible aerosol. Anecdotal evidence has shown that it causes milder throat hit than propylene glycol. Because of this, it is more commonly used in low- or non-nicotine liquids and facilitates the pattern of direct lung inhalation (with the user inhaling the aerosol directly to the lungs instead of keeping it in the oral cavity when taking the puff and subsequently inhaling it).

As with propylene glycol, glycerol exposure from e-Cigarette use is through inhalation of aerosol produced by heating. Limited evidence exists for this route of exposure. A study in Sprague-Dawley rats evaluated nose-only inhalation of glycerol aerosol for 2 and 13 weeks (5 days per week, 6 hours per day).[104] The mean exposure concentrations of the aerosol were 0, 7.00 ± 0.08, 1.93 ± 0.123, and 3.91 ± 0.458 mg glycerol/L of filtered room air. The mass median aerodynamic diameter (MMAD) was reported to be $<1.5 \, \mu m$. All rats underwent complete necropsy, and designated tissues were weighed and examined histopathologically, while blood was collected and analyzed for specific hematological and clinical chemistry parameters. The only major finding was minimal to mild squamous metaplasia of the epithelium lining at the base of the epiglottis. There was no effect on lung, liver, kidney, brain, and heart weight.

Current evidence is limited but does not raise concerns about serious adverse effects from the use of glycerol in e-Cigarettes. The systematic absorption of glycerol appears to be reasonably safe. The regulatory aspects are similar to the case of propylene glycol. It is important to implement the use of pharmaceutical grade glycerol in e-Cigarette liquids. Research on its safety for long-term inhalation should expand, while long-term epidemiological studies on e-Cigarette users will eventually evaluate its safety profile. Finally, the use of alternative compounds with similar properties (acting as a solvent and producing visible aerosol) should be evaluated.

Flavorings

Flavorings are an integral part of e-Cigarette liquids. e-Cigarettes are almost flavorless without the use of flavoring additives. A survey of 1347 e-Cigarette consumers found that only 1% used flavorless liquids.[105] Another survey of 4618 consumers found that, on average, they were using three different types of liquid flavors on a regular basis.[106] The majority were switching between flavors daily or even within the day, with former smokers switching more frequently than current smokers. Fruit flavors were more popular at the time of participation, while tobacco flavors were more popular at initiation of use. Consumers reported that flavors were very important in their effort to reduce or quit smoking, and the number of flavors regularly used was independently associated with smoking cessation.

There are mainly three types of food flavorings. Natural flavoring substances are obtained from plant or animal raw materials by physical, microbiological, or enzymatic processes. They can be used in their natural form or processed for consumption by humans. However, they cannot contain any nature-identical or artificial flavoring substances. Nature-identical flavoring substances are obtained by synthesis or are isolated through chemical processes. They are chemically identical to flavoring substances naturally present in products intended for consumption by humans. They cannot contain any artificial flavorings. Artificial flavoring substances are those that are not identified in a natural product intended for consumption by humans, whether or not the product is processed. They are typically produced by fractional distillation and additional chemical manipulation of naturally sourced chemicals or from crude oil or coal tar.

In 1959, The Flavor and Extract Manufacturers Association of the United States (FEMA) took its initial actions to establish a novel program to assess the safety and GRAS status of flavor ingredients as described in the 1958 Food Additives Amendments to the Federal Food, Drug, and Cosmetic Act, which is the federal law governing the regulation of flavors and other food ingredients.[107] Since then, FEMA GRAS has become the longest-running and most widely recognized industry GRAS assessment program. It began with a survey of the flavor industry to identify flavor ingredients then in use and to provide estimates of the amounts of these substances used to manufacture flavors. The initial FEMA Expert Panel was established in 1960, beginning its evaluation of the safety of flavor ingredients that continues today. The statutory definition of GRAS has four key criteria, all of which must be met for a food ingredient to be considered generally recognized as safe and exempt from the requirements for food additive approval: (1) There must be general recognition of safety by qualified experts. (2) The experts must be qualified by training and experience to evaluate the substance's safety. (3) The experts must base their determination of safety on scientific procedures or on common use in food prior to 1958. (4) The determination of general recognition of safety must take into account the conditions of intended use for the substance, in other words its function in the food, such as flavoring. A key part of the requirements for GRAS status is that there must be general recognition of safety and not just an assertion of safety per se. This means that there must be an opportunity for interested parties to review the basis for GRAS determinations, not just the conclusions. Specific criteria for the evaluation and establishment of GRAS status for flavoring additives have been developed.[108,109]

In the United States, FDA has the primary legal responsibility for determining safe use.[110] To market a new food or color additive, or before using an additive already approved for one use in another manner not yet approved, a manufacturer or other sponsor must first petition FDA for its approval. These petitions must provide evidence that the substance is safe for the ways in which it will be used. As a result of legislation in 1999, indirect additives have been approved via a premarket notification process requiring the same data as was previously required by petition. Under the Food Additives Amendment, two groups of ingredients were exempted from the regulation process. Group I (prior-sanctioned substances) are substances that FDA or USDA had determined safe for use in food prior to the 1958 amendment. Examples are sodium nitrite and potassium nitrite, used to preserve luncheon meats. Group II (GRAS) ingredients are those that are generally recognized by experts as safe, based on their extensive history of use in food before 1958 or on published scientific evidence. Among the several hundred GRAS substances are salt, sugar, spices, vitamins, and monosodium glutamate. Manufacturers may also request that FDA review the industry's determination of GRAS Status. If new evidence suggests that a product already in use may be unsafe, or if consumption levels have changed enough to require another look, federal authorities may prohibit its use or require further studies to determine if the use can still be considered safe. When evaluating the safety of a substance and whether it should be approved, FDA considers (1) the composition and properties of the substance, (2) the amount that would typically be consumed, (3) immediate and long-term health effects, and (4) various safety factors. The evaluation determines an appropriate level of use that includes a built-in safety margin, a factor that allows for uncertainty about the levels of consumption that are expected to be harmless. In other words, the levels of use that gain approval are much lower than what would be expected to have any adverse effect. Because of inherent limitations of science, FDA can never be absolutely certain of the absence of any risk from the use of any substance. Therefore, FDA must determine, based on the best science available, if there is a reasonable certainty of no harm to consumers when an additive is used as proposed.

In Europe, the risk assessment of flavoring substances has been carried out by the Panel on Food Contact Materials, Enzymes, Flavorings and Processing Aids (CEF), established by the European Food Safety Authority (EFSA), since 2008. Previously the task was done by the former Panel on Food Additives, Flavorings, Processing

Aids and Materials in Contact with Food (AFC). A flavoring is authorized following an application to the European Commission by an interested party, usually the producer of the flavoring. EFSA provides scientific advice on flavorings. It also provides the scientific basis of the program of evaluation of flavoring substances, guidelines on how to evaluate flavorings, and guidance on the data to be submitted for applications on flavorings. EFSA scientists look at intake levels, absorption, metabolism, and toxicity of individual substances within the group. Where EFSA identifies data gaps, it states the need for further data to the applicant and to the European Commission. In 1996, Regulation (EC) No. 2232/96 of the European Parliament and of the Council first laid down a Community procedure for flavoring substances used or intended for use in or on foodstuffs.[111] The criteria established to authorize flavoring substances were as follows: (1) They should present no risk to the health of the consumer. (2) Their use does not mislead the consumer. (3) Appropriate toxicological evaluation is needed to assess the possible harmful effects. (4) All flavoring substances must be constantly monitored and must be reevaluated whenever necessary. In 2010, EFSA published a document with guidance on the data required for the risk assessment of flavorings used in food products.[112]

A large number of flavors are available in the market. EU estimated that in 2012 there were 2800 compounds, of which 2100 had undergone evaluation and about 400 would finish the approval process by the end of 2015. In 2014, researchers identified more than 7000 flavored e-Cigarette liquids on the market.[113] Although the compounds used in the liquids are approved for use in food, their primary intended use (in food products) means that their safety was mainly assessed for ingestion only. FEMA released a statement in 2013 (updated in 2016) correctly saying that the GRAS provision applies only to food.[114] The fundamental differences between ingestion and inhalation are the direct exposure of the lungs, the fast absorption and delivery of the inhaled chemicals to the arterial system, and the bypass of the first pass effect by the liver. Additionally, the flavoring compounds are heated in the e-Cigarette device before being inhaled. Few flavoring substances have PELs promulgated by OSHA. PELs have the force of regulation and are an important way that OSHA regulates workplace safety. FEMA released a report in 2012 presenting a list of priority flavoring substances that are of concern for the respiratory health of workers in flavoring manufacturing facilities, and making suggestions on protective measures and label warnings.[115] The list included high and low priority compounds, based on available inhalation exposure data in animals and humans, chemical structure, volatility, and volume of use. Obviously, the list refers to occupational exposure, which is usually defined as continuous 8-hour daily exposure. This is different from the intermittent nature of e-Cigarette aerosol exposure and the relatively low number of puffs taken daily by consumers compared to the breathing rate over 8 hours in an occupational setting. It has been supported that the acceptable occupational exposure limits are not applicable to consumer exposure and should not be used in the case of e-Cigarettes.[114] However, this should be evaluated in the context of the intended purpose of e-Cigarette use, which is as a substitute for smoking. Smokers have a very high risk of developing respiratory disease, so it would be reasonable to accept the use of occupational safety limits in the risk assessment of an alternative-to-smoking product, especially when other relevant information is not available.[116] Although the lack of evidence on the safety for inhalation does not necessarily mean that these compounds are toxic, eventually their assessment for the safety of inhalation is necessary. This will be an expensive task and will take a long time to be completed.

Many years will be required to scientifically establish the safety of all the flavoring compounds for inhalation, especially for local effects on lung tissue. At the same time, flavorings are important in the acceptance of e-Cigarettes among smokers and in their efficacy as smoking substitutes.[106] Thus the restriction of the use of flavors until they are proven absolutely safe would almost certainly result in the elimination of e-Cigarettes. This is a complex regulatory issue. The EU recently introduced legislation for e-Cigarettes, which does not apply restrictions to the use of flavoring compounds in the liquids but gives to the members states the responsibility for adopting relevant rules.[9] It also requires that only ingredients of high purity that do not pose a risk to human health should be used. At the same time, it states that if some products are found to pose a serious risk to human health, appropriate provisional measures should be taken, including the withdrawal of the products. This is a reasonable approach, considering the above-mentioned limitations of the high cost and the long time needed to establish the safety of flavors for inhalation.

FLAVORING COMPOUNDS OF POSSIBLE REGULATORY CONCERN FOR e-CIGARETTES

It is beyond the scope of the chapter to evaluate separately the evidence on the safety for inhalation of each flavoring compound, considering the very large number available and the lack of data identifying all compounds that are currently used in e-Cigarette liquids. However, it is important to present data on two chemicals which have been found in e-Cigarette liquids and could be a reason for concern in terms of respiratory adverse effects.

Diacetyl (CAS Reg. No. 431-03-8, Fig. 5.4A), also known as 2,3-butanedione, is a member of a general class of organic compounds referred to as diketones, α-diketones, or α-dicarbonyls. It is a volatile green to yellowish liquid with a boiling point of 88°C.[117] It has an odor threshold concentration of 0.05−4 µg/L in water[117,118] and 0.01−0.02 ppb in air.[119] Diacetyl is responsible for providing a characteristic buttery and creamy flavor, and is both naturally found in food and used as a synthetic flavoring agent in food products such as butter, caramel, cocoa, coffee, dairy products, and alcoholic beverages.[120] It is approved for use as a food flavoring additive (FEMA No. 2379). It is also formed endogenously in humans, from decarboxylation of pyruvate.[121] Synthetically, it is produced in different ways:[117,121] (1) from methyl ethyl ketone, either by converting it to an isonitroso compound and then hydrolyzing with hydrochloric acid or by partial oxidation of methyl ethyl ketone over a copper or vanadium oxide catalyst; (2) from 2,3-butanediol, by oxidative dehydrogenation of 2,3-butanediol over a copper or silver catalyst; (3) from acetoin (obtained by electrochemical oxidation of methyl ethyl ketone), by reacting acetoin with molecular oxygen in the presence of copper oxide catalyst; (4) from 1-hydroxyacetone, by the acid-catalyzed condensation of 1-hydroxyacetone with formaldehyde. Diacetyl is also a byproduct of fermentation. Natural diacetyl is used in the form of starter distillate,

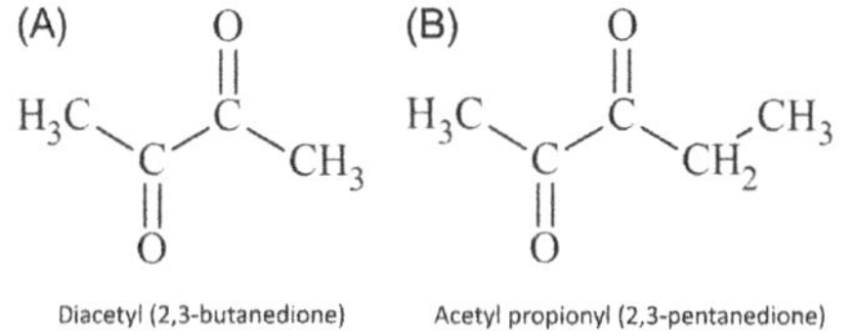

Figure 5.4 Chemical structures of diacetyl (A) and acetyl propionyl (B).

a concentrated flavor distillate, which may contain different concentrations of diacetyl depending on production conditions.[117] In mammalian cells, diacetyl is metabolized principally by reduction to acetoin, a reaction catalyzed by dicarbonyl/L-xylulose reductase (diacetyl reductase).[122–124] Diacetyl reductase is present in the rat liver, kidney, brain,[117,122] and respiratory mucosa, especially in the olfactory epithelium.[125] Other metabolic pathways have also been found in the rat and human lungs.[126,127]

Acetyl propionyl (CAS Reg. No. 600-14-6, Fig. 5.4B), also known as 2,3-pentanedione, is similar to diacetyl and also belongs to the organic compounds class of α-diketones. It is a volatile yellow liquid with a boiling point of 108°C.[117] It has an odor threshold concentration of 30 µg/L in water,[117] and 0.01−0.02 ppb in air.[119] It has similar flavoring properties and characteristics to diacetyl and is approved for use as a flavoring additive in food (FEMA No. 2841). It also occurs naturally in many ingredients, including some essential oils, nuts, red meat, poultry, seafood, alcoholic beverages, and fruits. Naturally, it is produced by fermentation.[116] Synthetically, it can be produced by[117] (1) the condensation of lactic acid and an alkali metal lactate, (2) the acid-catalyzed condensation of 1-hydroxyacetone with paraldehyde, or (3) the oxidation of 2-pentanone with excess sodium nitrite and diluted hydrochloric acid in the presence of hydroxylamine hydrochloride. Similarly to diacetyl, it is metabolized in mammalian cells by diacetyl reductase.[122]

Diacetyl and acetyl propionyl have similar uses in the food industry and are therefore discussed jointly. They represent a typical case of compounds that have GRAS status for ingestion but for which there is evidence of potential toxicity to the lungs from inhalation. Diacetyl exposure has been linked to the development of bronchiolitis obliterans, a rare and life-threatening form of nonreversible obstructive lung disease in which the bronchioles are compressed and narrowed by fibrosis and inflammation. It is characterized by decreased FEV_1 and FEV_1/FVC ratio on spirometry testing, and is associated with varying degrees of shortness of breath. Pathologically, there are alterations in the walls of respiratory and membranous bronchioles that cause concentric narrowing or complete obliteration of the airway lumen, without involvement of the distal lung parenchyma by inflammation or organizing pneumonia. This condition is most commonly observed as a manifestation of chronic lung allograft dysfunction after lung transplantation.[128] The definite diagnosis of the condition is done by lung biopsy, which may be difficult due to its patchy distribution;[129,130] thus the condition may be often misdiagnosed.[131] It has also been observed after lung infections[132,133] and after exposure to sulfur mustard.[134]

The first cases of fixed obstructive lung disease suggestive of bronchiolitis obliterans were observed in two workers in a facility where flavorings with diacetyl were made for the baking industry.[135] In May 2000, eight persons who had formerly worked at a plant that produced microwave popcorn were reported to the Missouri Department of Health to have bronchiolitis obliterans.[136] Kreiss et al. first identified the link between exposure to diacetyl and the risk of lung disease in 2002.[131] They performed a cross-sectional medical and environmental survey at the index plant in November 2000 and linked this outbreak to inhalation exposure to butter flavorings used in the production process. The finding of excess airway obstruction among current workers included a striking excess among quality control (QC) workers who performed microwave popping of approximately 100 bags of product per worker per work shift. Five of six workers had airways obstruction in this setting. The observation of this case series led to the identification of another case of bronchiolitis obliterans in a separate facility, in a study by Parmet.[137] Since then, numerous cases of respiratory dysfunction after exposure to buttery flavorings have been observed in the microwave popcorn industry.[138–141] Additionally, a retrospective epidemiologic study found cases of bronchiolitis obliterans in workers who were employed in a chemical plant with exposures to diacetyl and other related flavoring compounds,[142] while another study identified increased respiratory disease-associated mortality rate among workers at the microwave popcorn company, especially for those employed before the company reduced diacetyl exposure.[143] Due to the increased prevalence of this disease in popcorn manufacturing facilities, the condition has been called "popcorn worker's lung." It should be mentioned that most cases were identified by clinical criteria and were not verified by biopsy. However, the studies indicate a higher prevalence of respiratory dysfunction and disease in workers exposed to buttery flavors.

After identification of cases of respiratory dysfunction among workers, experimental studies were done on animals to assess the effects of exposure to diacetyl, focusing on the upper and lower respiratory tracts. In rats, acute exposures to diacetyl or diacetyl-containing butter flavoring vapors caused necrosis in the epithelial lining of nasal and pulmonary airways. Hubbs et al. found necrotizing rhinitis in rats inhaling buttery flavoring vapors.[144] Another study by the same group identified epithelial necrosis and inflammation in bronchi, trachea, and larynx of Sprague-Dawley rats exposed to diacetyl.[145] Similar observations were made in mice.[146] One study was able to cause bronchiolitis obliterans by intratracheal instillation of diacetyl, but this was associated with the delivery of a very large single dose.[147]

Acetyl propionyl was investigated in animal experiments because it is structurally similar to diacetyl, being a 5-carbon α-diketone. Moreover, this flavoring has replaced diacetyl in the food flavoring industry after concerns were raised for the latter being associated with the development of lung disease.[148] One study found increased airway reactivity in an isolated perfused trachea preparation after exposure to acetyl propionyl, with the effects being more pronounced compared to exposure to diacetyl.[149] Bronchial fibrosis was observed in a study in rats,[150] and another study found similar fibrotic effects for both diacetyl and acetyl propionyl at similar levels of exposure.[151] A study by Hubbs et al. exposed rats to inhaled acetyl propionyl and found evidence of necrotizing rhinitis, tracheitis, and bronchitis.[152]

The data implicating diacetyl and acetyl propionyl to respiratory disease has led many regulatory organizations to propose acceptable occupational exposure limits (Table 5.1). In 2011, the

Table 5.1 Occupational Exposure Limits for Diacetyl and Acetyl Propionyl Proposed by Various Organizations			
Organizations	Exposure Time	Acetyl Propionyl	Diacetyl
ACGIH	15 min STEL	Not available	0.02 ppm (70 μg/m^3)
	8 h TWA	Not available	0.01 ppm (35 μg/m^3)
European Commission	15 min STEL	Not available	0.1 ppm (350 μg/m^3)
	8 h TWA	Not available	0.02 ppm (70 μg/m^3)
OSHA	15 min STEL	Not available	Not available
	8 h TWA	Not available	Not available
NIOSH	15 min STEL	0.031 ppm (127 μg/m^3)	0.025 ppm (88 μg/m^3)
	8 h TWA	0.0093 ppm (38 μg/m^3)	0.005 ppm (18 μg/m^3)

National Institute for Occupational Safety and Health (NIOSH) proposed 15-minute short-term exposure limits (STELs) of 0.025 ppm (88 μg/m^3) and 0.031 ppm (127 μg/m^3) for diacetyl and acetyl propionyl, respectively.[117] The 8-hour time-weighted average (TWA) Recommended Exposure Limits (RELs) were set at 0.005 ppm (18 μg/m^3) and 0.0093 ppm (38 μg/m^3), respectively. NIOSH estimated that this would be associated with a 1 in 1000 excess prevalence of pulmonary dysfunction after 45 years of exposure. More recently, the Scientific Committee on Occupational Exposure Limits (SCOEL) of the European Commission published a draft recommending diacetyl occupational exposure limits of 0.10 ppm (350 μg/m^3) as a 15-minute STEL and 0.020 ppm (70 μg/m^3) as an 8-hour TWA.[153] SCOEL did not follow the NIOSH approach because of concerns about the robustness of the exposure data on which the assessment was based. In 2012, ACGIH adopted TLVs for diacetyl, including a 15-minute STEL of 0.02 ppm (70 μg/m^3) and an 8-hour TWA of 0.01 ppm (35 μg/m^3).[154] No OELs have been defined by SCOEL and ACGIH for acetyl propionyl, and OSHA has not defined OELs for either flavoring.

Diacetyl and acetyl propionyl are also emitted in tobacco cigarette smoke.[155,156] The average levels reported in a recent study were 285 μg/cigarette for diacetyl and 43 μg/cigarette for acetyl propionyl, using the conservative ISO 3308 puffing regime.[157] The levels were much higher (804 and 127 μg/cigarette, respectively) when using the more realistic Health Canada intense puffing regime.[157] Interestingly, the authors commented that the major source of diacetyl and acetyl propionyl is the pyrolysis of the cigarette and not the use of these compounds as flavoring additives. This was verified by the finding of both compounds in the standardized 3R4F cigarettes, which contain only tobacco and humectants but not flavoring chemicals. Thus exposure to diacetyl and acetyl propionyl from smoking is unavoidable and would not be affected by changing the composition of the additives to the tobacco cigarette.

Two studies have systematically evaluated the presence of diacetyl and acetyl propionyl in e-Cigarette liquids. In 2014, a study examined 159 liquids from 36 manufacturers and retailers in the United States and Europe for the presence of the two flavoring compounds.[158] Both refill liquids ("ready to use") and concentrated flavors, which are diluted by users in base liquids (mixtures of propylene glycol, glycerol, and nicotine), were obtained. The authors specifically assessed sweet-flavored liquids, assuming that the buttery characteristics of diacetyl and acetyl propionyl would be more commonly used in such liquids. The majority of the samples (74.2%) contained at least one of the compounds. Diacetyl was found in 110 (69.2%) samples, with a median concentration of 29 μg/mL (IQR 10−170 μg/mL). The highest levels found were 32,115 μg/mL in flavor concentrates and 10,620 μg/mL in ready to use liquids. Acetyl propionyl was found in 53 (33.3%) samples, containing a median concentration of 44 μg/mL (IQR 7−172 μg/mL), with the highest levels being 3082 μg/mL in flavor concentrates and 1018 μg/mL in ready to use liquids. The authors calculated that the median daily exposure level was 56 μg (IQR 26 to 278 μg/day) for diacetyl and 91 μg (IQR 20 to 432 μg/day) for acetyl propionyl (Fig. 5.5). They tried to assess the level of exposure based on the strictest, NIOSH-proposed safety limits for occupational exposure. They calculated the 8-hour total exposure to diacetyl and acetyl propionyl by multiplying the NIOSH 8-hour TWA limits with a resting respiratory rate of 3.6 m^3/8 hours. With this method, they defined that the total amount of diacetyl that can be inhaled daily (according to NIOSH limits) was 65 μg (18 μg/m^3 × 3.6 m^3), while for acetyl propionyl it was 137 μg (38 μg/m^3 × 3.6 m^3). The median daily exposure level from using diacetyl-containing liquids was lower than the NIOSH-defined safety limit, but 52 samples (47.3% of the positive samples) would expose consumers to levels higher than the NIOSH limits. The sample with the highest level of diacetyl would result in a 490-fold higher daily intake than the NIOSH limit. For acetyl propionyl, the

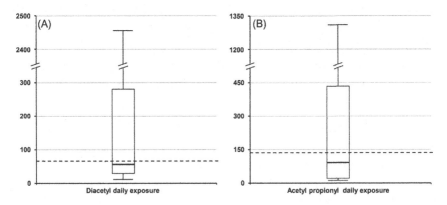

Figure 5.5 Estimated daily exposure to diacetyl (A) and acetyl propionyl (B) from a study evaluating their levels in 159 e-Cigarette liquid samples. Reproduced with permission from Farsalinos et al.[158]

median level of daily exposure was also lower than the NIOSH-defined safety limit (137 μg/day); however, 22 samples (41.5% of the positive samples) would expose consumers to levels higher than the NIOSH limit. The sample with the highest level of acetyl propionyl would result in a 22-fold higher daily intake than the NIOSH limit. Of note, their method was underestimating the NIOSH-defined total daily exposure safety limits because of using the resting respiratory rate. By using a more realistic respiratory rate related to the occupational setting, they found that the acceptable daily exposure limit would be 86 μg/day for 8 hours of light activity and 302 μg/day for 8 hours of moderate activity for diacetyl, and 182 and 638 μg/day respectively for acetyl propionyl.[116] An additional risk assessment analysis was performed by comparing the levels found in liquids to the levels of exposure from smoking 20 cigarettes per day. They found that smoking would result in 100-fold higher exposure to diacetyl and 10-fold higher exposure to acetyl propionyl compared to e-Cigarette use. Finally, they analyzed the presence of diacetyl and acetyl propionyl in e-Cigarette aerosol by preparing three samples with known concentrations and collecting the aerosol from 50 puffs. Similar concentrations of diacetyl and acetyl propionyl were found in the liquids and the respective aerosol collection samples, while a very strong correlation was observed between the expected (based on the liquid consumption) and observed (measured) concentrations ($R^2 = 0.997$ for diacetyl and $R^2 = 0.995$ for acetyl propionyl). The results indicated that both diacetyl and acetyl propionyl are readily delivered from the liquid to the aerosol, which is an expected finding due to the high volatility of these compounds. The authors concluded that, since these chemicals are not formed from the heating process, they represent an unnecessary and avoidable risk. They proposed that manufacturers and flavoring suppliers should take the necessary steps to make sure that these chemicals are not present in e-Cigarette liquid products, by regularly testing their products and changing formulations, without the need to limit the availability of sweet flavors in the market. The use of the occupational recommended exposure limits in this study was criticized by NIOSH in a letter to the editor, suggesting that these were not intended to establish safe exposure concentrations for consumers or the general public.[159] The authors of the original study acknowledged this fact but mentioned that smokers are a specific population group exposed to a large number of toxins on a daily basis, irrespective of being healthy and fit to work or not.[116] Smokers have a very high risk of developing lung disease; thus, if e-Cigarettes are used as smoking substitutes, the risk associated with exposure to diacetyl and acetyl propionyl would be minimal compared to continuous smoking. Moreover, exposure from e-Cigarettes is intermittent, only when taking a puff, while the occupation exposure limits are related to continuous (with every breath) exposure for 8 hours daily.

A more recent study evaluated the presence of diacetyl and acetyl propionyl in 51 e-Cigarette products.[160] Aerosol collections and measurements were performed, but the samples were collected until the e-Cigarettes were exhausted, determined by the lack of visible emissions in the chamber. This could result in significant overheating, with a resulting overestimation of exposure compared to realistic use patterns. Diacetyl was found to be above the limit of detection in 39 of the 51 flavors tested and above the limit of quantification in 34 flavors. The levels ranged from 0.3 to 239 μg/e-Cigarette. Acetyl propionyl was detected in 23 flavors above the limit of detection and in 21 flavors above the limit of quantification. The levels ranged from 0.2 to 64 μg/e-Cigarette. Twenty-one flavors contained both diacetyl and acetyl propionyl above the limit of detection. On average, the levels were much lower compared to the previous study, although the reporting values were not expressed per liquid volume or aerosol yield. The authors did not present any risk assessment analysis, but stated that they disagree with the use of occupational exposure limits as a method of identifying potential risks. Also, they did not comment on the abundant presence of the two flavoring compounds in tobacco cigarette smoke.

Some scientists dispute the association between diacetyl exposure and development of bronchiolitis obliterans. Pierce et al. argued that, although cigarette smoking has been linked to numerous respiratory diseases, there are no cases of bronchiolitis obliterans that have been attributed to cigarette smoking.[157] Considering their findings that diacetyl levels in tobacco cigarette smoke are almost always higher than levels associated with workplace settings, they suggested that compounds other than diacetyl may be implicated in the development of bronchiolitis obliterans among workers in flavoring manufacturing facilities. Kreiss et al. found a significantly higher prevalence of airway obstruction among nonsmokers compared to that of smokers,[131] while Kanwal et al. found that nonsmokers had a higher prevalence of obstruction and a lower FEV$_1$ than smokers.[161] These observations suggest a potential protective effect of smoking.[137] Although this phenomenon needs to be further examined, it should be noted that cigarette smoke contains many respiratory toxins, which probably act synergistically and cause a different pattern of lung disease. The prevalence of chronic obstructive lung disease in active smokers is estimated to be 15.4%,[162] far higher than the prevalence of bronchiolitis obliterans in patients exposed to diacetyl. Moreover, the condition is often misdiagnosed. Finally, postmortem examinations have shown that many smokers have histopathological features of respiratory bronchiolitis.[163]

In conclusion, there is reason for concern about the risk of developing respiratory disease from exposure to diacetyl and acetyl propionyl. In tobacco cigarettes, exposure to these chemicals is an unavoidable risk because they are produced by pyrolysis. Contrary to that, in e-Cigarettes the source of the compounds is their introduction as flavoring additives during the production process of e-Cigarette liquids. Production of diacetyl and acetyl propionyl from thermal decomposition seems to be unlikely, based on a study which compared their levels in the liquid and the aerosol.[158] Thus they represent an unnecessary and avoidable risk. Obviously, more studies are needed to establish a clear link between diacetyl and especially acetyl propionyl and respiratory damage. Currently, there are no reported cases of respiratory disease linked to exposure from e-Cigarettes. However, these chemicals could be a reasonable target for regulatory action. An attempt to introduce safety limits would be inapplicable and unrealistic because exposure is mostly dependent on daily liquid consumption, which differs greatly among consumers.

e-CIGARETTE EMISSIONS AS A REGULATORY ISSUE

Regulatory intervention related to e-Cigarette emissions could have a dual effect: control of nicotine delivery to the consumer and reduction of the amount of toxic emissions in the aerosol. e-Cigarettes are very diverse products, with different and extensive capabilities in aerosol production and delivery to the user. A recent study identified large differences in nicotine delivery potential between first and new generation devices.[164] In many cases, nicotine yield exceeded that of tobacco cigarettes, although different puffing patterns from those usually employed in tobacco cigarette studies were used. Additionally, different atomizers had to be used at specific power settings (watt delivery from the battery) in order to be functional and represent realistic use. This is related to the design characteristics of the atomizer and the energy that can be delivered from the battery. Users have an unprecedented ability to control operating conditions and device performance in order to get a satisfactory experience. Talih et al. evaluated aerosol yield and nicotine delivery in a laboratory setting by adjusting puffing patterns and voltage delivery to the atomizer and collecting aerosol samples.[165] They found that voltage setting and puff duration were the major determinants of aerosol and nicotine yield, while puff volume had no effect. Another study found that plasma nicotine delivery was directly related to nicotine content of the liquid.[166] Based on these, it would be tempting to regulate nicotine delivery by setting specific standards in nicotine concentration of the liquids and functional and performance characteristics of the devices. In fact, Shihadeh and Eissenberg proposed that nicotine delivery should be regulated in terms of total dose and rate of nicotine delivery.[167] The purpose would be to address the issue of drug abuse liability, which is well known to be affected by both factors.[168,169] The proposal included the definition of an acceptable range of nicotine delivery rates and levels and measures to adjust the performance characteristics of the devices and the nicotine content of the liquids accordingly in order not to exceed the proposed limits. This proposal is very complex, due to the many different factors that need to be controlled. The approach has been challenged and criticized.[170] The critics mentioned that, from a safety perspective, there is no risk of acute nicotine intoxication, because the phenomenon of self-titration of nicotine intake is well established from tobacco cigarette research. Of note, the same phenomenon has been recently verified in e-Cigarettes.[171] Also, the restriction of nicotine delivery would result in less appealing products and lower efficacy of e-Cigarettes as smoking substitutes. Although perceived pleasure should not be a reason for anyone to initiate smoking (due to exposure to known risks for health damage) or e-Cigarette use (due to exposure to yet unknown, although evidently much lower than smoking, risk), it should be considered and evaluated when discussing smoking substitution. The enormous variability in e-Cigarette devices and liquids is explained by the main criterion of choice for consumers, which is self-preference.[72,106] Consumers have different patterns of e-Cigarette use. They can consume the same amount of liquid by either taking more puffs with lower duration or fewer puffs with higher duration, depending on perceived satisfaction. They can adjust the selection of devices and liquids so as to get the pleasure they expect in terms of patterns of use and nicotine intake. Restricting the performance characteristics by introducing power or puff duration cut-off would deprive the consumer of use according to their preferred pattern. That would give e-Cigarettes characteristics similar to medications, which by definition are associated with strictly controlled and specific delivery of dosage and use patterns. Finally, another important issue is whether it is appropriate or even ethical to apply restrictions to e-Cigarettes, which could result in reduced efficacy as a smoking substitute and thus cause harm to smokers, because another part of the population (nonsmokers) could decide to voluntarily adopt its use and expose themselves to a new (even minor) risk.[170] This is a controversial issue that forms most of the philosophical discussion about accepting or rejecting a harm reduction approach and principle. There are other practical issues with regulation of nicotine delivery. The e-Cigarette market is mostly characterized by liquids, batteries, and atomizers being sold separately. This results to innumerable combinations. Thus it is extremely difficult to control or implement specific combinations of products unless all these products are banned and only fixed combinations are allowed. The EU's recent Tobacco Products Directive introduced a maximum level of 20 mg/mL nicotine concentration in e-Cigarette liquids.[9] However, this decision was not based on an attempt to regulate nicotine delivery to e-Cigarette users but was related to the reduction of the risk of accidental, unintended exposure (e.g., ingestion of liquids by children). For the same reason, a limit of 10-mL volume for nicotine-containing refill liquids has been implemented. The approach to regulating nicotine delivery is a challenging, complex, and difficult task that would probably require the complete transformation of the market and could have unintended negative consequences, mainly related to reduced appeal and use of the products as tobacco substitutes. It is almost certain that the debate on this issue will continue for a long time, with different approaches and opinions expressed by the scientific community. Obviously, it is reasonable for the public health community to set a goal of reducing, and to discourage, nicotine use; however, the most important priority should be to prevent disease and death, and to achieve this to the highest level possible it might be more appropriate to consider the elimination of nicotine use as a secondary goal.

A necessary and reasonable regulatory decision would be to implement the mandatory use of pharmaceutical grade nicotine in e-Cigarettes. Commercially available nicotine in all products, including pharmaceutical nicotine replacement therapies, is extracted from tobacco. The mandatory use of high-purity (pharmaceutical grade) nicotine would ensure that impurities extracted from tobacco, such as the known carcinogens tobacco-specific nitrosamines, are present at minimum levels. The European Pharmacopeia states that pharmaceutical grade nicotine may contain up to 0.3% of each of the specified nicotine impurities (the minor tobacco alkaloids anatabine, beta-nicotyrine, cotinine, myosmine, nicotine-N-oxide, nornicotine, and anabasine), 0.1% each of unspecified impurities, and total amount of less than 0.8% for all impurities.[172] The United States Pharmacopeia states that no single impurity should be more than 0.5% and that total amount of impurities should be less than 1.0%.[173] Etter et al. analyzed 20 e-Cigarette liquid samples from 10 brands for the presence of these impurities.[174] Most products contained some, with levels up to 4.4%. In most cases, the levels were between 1% and 2%. Another study by the US Centers for Disease Control and Prevention analyzed 36 samples from 4 manufacturers for the presence of nicotine impurities.[175] All samples contained some but, for the majority, the levels were below the USP limits. Three samples contained one impurity at levels exceeding the acceptable limits and two of them also exceeded the limit for the total level of impurities. The presence of impurities found in these studies does not necessarily mean that low-purity nicotine was used in the production of e-Cigarette liquids. Oxidative degradation of nicotine can result in the formation of minor tobacco alkaloids.[176,177] This can happen either during the manufacturing of the ingredient or during the manufacturing of the final liquids. Also, there may be some interactions with the packaging material and the flavoring compounds, or inadequate handling and storage conditions.[174] Oxidation can happen during the shelf life of the product because of the presence of air inside the liquid container. The rate of oxidation of nicotine-containing e-Cigarette liquids is currently unknown, thus the source of impurities cannot be identified with certainty.[175]

Another area of regulatory interest is the restriction on the use of substances that could potentiate the addictiveness of e-Cigarettes. It is well established that compounds such as acetaldehyde and chemicals inhibiting monoamine oxidase, which are naturally present in tobacco or are produced from pyrolysis, can potentiate the addictiveness of nicotine.[178-182] β-carboline alkaloids such as harman and norharman have been identified in tobacco leaves and cigarette smoke and have monoamine oxidase inhibitor properties.[183] Such compounds are also present in food products, especially in processed food.[184] There is some evidence that minor tobacco alkaloids may have similar effects,[185] although other studies have disputed that.[186] Other additives may interfere with nicotine kinetics by altering the absorption and elimination of nicotine. One such example is nicotyrine. This minor tobacco alkaloid is a potent reversible inhibitor of human cytochrome P450 2A13 (CYP2A13) in the nasal and respiratory epithelium and an irreversible inhibitor of CYP2A6 in the liver.[187-190] These enzymes are responsible for the metabolism of nicotine in humans. It has been proposed that e-Cigarettes may be more effective in nicotine delivery due to the presence of nicotyrine, which facilitates absorption and slows hepatic clearance of nicotine,[190] but there is not enough evidence yet to substantiate this hypothesis. However, it could be tempting for manufacturers to introduce to e-Cigarette liquid compounds that potentiate nicotine. That could have a dual effect: allow e-Cigarettes to deliver a better nicotine experience to smokers, which might enhance their effectiveness in smoking substitution, and make e-Cigarettes more addictive. Therefore, this is a controversial issue, given the potential benefits and harms. There are already anecdotal reports that e-Cigarette liquids containing monoamine oxidase inhibitors are available in the market, and that they are produced through a specific extraction process from tobacco and not added as raw ingredients. There is currently no evidence that such liquids are better in relieving nicotine craving or in promoting smoking abstinence, and it is not known if the specific extraction process introduces more contaminants to the liquids. A regulatory decision to use pharmaceutical grade nicotine would preclude the unintended introduction of these chemicals to the liquids. As for the intentional addition of nicotine-potentiating chemicals, that would depend on the balance between increased effectiveness of e-Cigarettes as smoking substitutes and increased abuse potential by nonsmokers. It would be reasonable to implement restrictions on the use of such compounds until more evidence determines their effects.

e-Cigarettes emit to the aerosol potentially harmful chemicals such as aldehydes, namely formaldehyde, acetaldehyde and acrolein, derived from thermal degradation of e-Cigarette liquid components.[191,192]

The issue of aerosol emissions is covered in depth in Chapter 2. From a regulatory perspective, it would be tempting to restrict the performance characteristics of e-Cigarette devices in an attempt to reduce toxic emissions. Studies have shown that aldehyde emissions are related to the power delivered to the atomizer.[193–195] Thus one possible decision would be to limit power delivery. However, this is an oversimplified approach. Heat generation and aerosol production depend on the total energy delivered to the atomizer.[194] The energy unit is the joule (J), which is defined as power (W) multiplied by time (s). W is defined as the square of voltage (V^2) divided by resistance (R). A major problem observed in many but not all studies is that they report V rather than W levels.[193,196] This should be avoided because the energy (and thus heat generation) is inversely related to the resistance value of the atomizer. When using similar V levels to atomizers of different resistance values, there are large discrepancies in power and energy as well as heat generation and aerosol yield. Therefore, to be able to compare different use conditions and equipment characteristics, it is appropriate to report power level (W) and time of applying the power. Another issue, infrequently addressed in laboratory studies evaluating aerosol emissions, is the dry puff phenomenon. This is an unpleasant taste, due to liquid overheating, that the user detects and avoids by reducing power level and puff duration or by increasing interpuff interval. This phenomenon has been described by users in internet forums and was first mentioned in the literature in 2013.[197,198] It is an organoleptic parameter and thus a subjective sign of overheating. In a study evaluating e-Cigarette use topography, researchers had to exchange the atomizer provided to the participants with another model because some users complained about dry puffs preventing them from using the e-Cigarette ad libitum.[198] Studies have shown that when the energy delivered to the atomizer is high enough to generate dry puffs, the level of aldehyde emissions is substantially increased.[194] Therefore, it actually represents a natural defense mechanism against exposure to overheated liquid emissions. Testing e-Cigarette aerosol chemistry in a laboratory setting without identifying dry puff conditions can result in overestimation of aldehyde emissions, which would represent unrealistic exposure.[194,199] At this time, it is unknown whether this phenomenon is consistent among different users; although every e-Cigarette user knows about its existence, there is no research evaluating the inter-individual differences in the power settings and energy delivery needed to detect dry puffs. The difference between atomizer design and performance characteristics means that each atomizer can be used at different energy levels before generating dry puffs. There is no specific energy level at which the dry puff phenomenon is generated in all available atomizers. So a regulatory decision to limit energy delivery from e-Cigarettes would require separate levels set for the innumerable combinations of battery and atomizer products. Moreover, there is an established compensatory pattern of e-Cigarette use when different conditions are used. At an FDA workshop on e-Cigarettes, a study was presented showing that in a clinical setting of e-Cigarette use with 30-minute sessions, the puff duration decreased by 0.9 seconds on average when the power setting was changed from 6 W to 10 W.[200] Puff number also decreased but liquid consumption was slightly increased. This is evidence that e-Cigarette users adjust the puffing pattern to generate the aerosol yield they prefer and need. Even if the devices were regulated to provide specific energy levels, still consumers would increase the number of puffs. Such a regulatory approach would be unrealistic and unfeasible given the current marketing situation and device availability.

A new technology that could prove valuable in reducing toxic emissions from thermal degradation is the ability to adjust and limit the temperature of the heating coil. This temperature control characteristic is based on the principle that some metals change their resistivity linearly depending on their temperature due to a higher number of collisions between ions inside the wire when the temperature is elevated.[187] The resistance at temperature t (R_t) can be calculated by the equation:

$$R_t = R_0 \times [1 + a(t - t_0)]$$

where R_0 is the resistance at a baseline temperature, α is the temperature coefficient of resistance of the material (expressed in $°C^{-1}$), and t_0 is the baseline temperature. Temperature coefficients of resistance for various materials have been reported, usually in websites (Table 5.2).[201,202] The devices with temperature control capability include electronics that detect the resistance of the coil and allow users to set the preferred maximum temperature of the coil. Usually the baseline temperature is measured by a sensor inside the battery device. Instructions on such devices usually report that the baseline resistance should be set with the coil at room temperature, so that the correct temperature related to the baseline resistance value is determined. Then, by continuously measuring the resistance when the device is

Table 5.2 Temperature Coefficients of Resistance Reported on Websites for Several Materials	
Materials	Temperature Coefficient of Resistance (α)
Aluminum	4.29×10^{-3}
Carbon (graphite)	-5×10^{-4}
Copper	3.86×10^{-3}
Copper, annealed	3.93×10^{-3}
Germanium*	-5×10^{-7}
Gold	3.4×10^{-3}
Iron	6.51×10^{-3}
Manganin	2×10^{-6}
Mercury	9×10^{-4}
Nichrome (Ni–Fe–Cr alloy)	4×10^{-4}
Platinum	3.927×10^{-3}
Silver	3.8×10^{-3}
Tungsten	4.5×10^{-3}
Materials Commonly Used in e-Cigarettes	
Nickel Ni200	6×10^{-3}
Stainless steel 304	1.016×10^{-3}
Stainless steel 316	0.88×10^{-3}
Stainless steel 321	0.917×10^{-3}
Stainless steel 340	1.38×10^{-3}
Titanium	3.66×10^{-3}
Information was derived from references 187 and 188.	

activated and applying the above formula, the temperature during e-Cigarette activation can be estimated. Through specific algorithms in the firmware, the power delivery of the battery device is adjusted so that the resistance does not exceed the calculated value related to the temperature set by the user. This technology could be implemented as mandatory by a regulatory decision. However, presently there is no research defining the temperature range associated with meaningful reductions in toxic emissions and whether this approach is more effective in reducing exposure to toxins than the subjective identification of dry puffs.

Another issue that has generated little research interest until now is metal emissions. e-Cigarette atomizers are mainly metallic, thus some metals may be emitted to the aerosol. Goniewicz et al. measured heavy metal emissions to the aerosol generated from 12 e-Cigarette products.[203] The following heavy metals were analyzed: arsenic, barium, cadmium, chromium, cobalt, copper, lead, manganese, nickel, rubidium, strontium, and zinc. The findings were reported as amount per 150 puffs. Only three metals—cadmium, lead, and nickel—were detected at levels above the limit of quantification. Overall, the levels were very low and were also detected in blank samples and in a pharmaceutical nicotine inhaler, raising the possibility that the environmental air was the source. Williams et al. evaluated the presence of several heavy metals in the aerosol emitted from one cartomizer and reported the results as amount per 10 puffs.[204] The following were detected above the limit of quantification: aluminum, barium, chromium, copper, iron, lead, manganese, nickel, strontium, tin, titanium, zinc, and zirconium. The authors did not report whether metals were measured in the environment. A risk assessment analysis was performed, evaluating the metal emissions from both studies compared to exposure limits defined by various organizations.[205] Specifically, Permissible Daily Exposure (PDE) limits for inhalational medications defined by the United States Pharmacopeia, Minimal Risk Levels (MRLs) defined by the Agency for Toxic Substances and Disease Registry, and Recommended Exposure Limits (RELs) defined by NIOSH were used depending on data availability. The analysis was based on a presumed daily consumption of 1200 puffs of e-Cigarette, which the authors considered as double the usual consumption. For MRLs and RELs, the average acceptable daily exposure was measured based on the amount of air inhaled over a period of 24 hours and 10 hours, respectively. For PDE limits, no calculation was needed since they refer to daily exposure. The average daily exposure from the 13 e-Cigarette products analyzed was 2.6- to 387-fold lower than the safety limits of PDE, 325-fold lower than MRLs, and 665- to 77,514-fold lower than RELs. One of the 13 products was found to result in exposure 10% higher than PDE for cadmium at the presumed daily use of 1200 puffs. The authors noted that, although the average level of metal exposure does not justify serious health concerns, significant differences in emissions were observed between different products, especially for metals of high concern such as cadmium, lead, and nickel. Thus there is a large margin for improvement by implementing strict quality standards and choosing appropriate materials. A more recent study evaluated four e-Cigarette products and found high emissions of tin in the products using friable solder joints in the wires inside the atomizers.[206] The authors proposed product modifications that could reduce tin emissions, such as replacing wire-to-wire tin solder joints with brass clamps, which are inherently more stable.

Metal emission measurements could be implemented through regulation, at least for metals of major concern. In the EU, regulation mandates the measurement and reporting of at least five metals in e-Cigarette emissions in order for a product to be approved and marketed.[207] Future research should define materials which are more stable when in contact with liquids and under the repeated heating and cooling cycles of e-Cigarettes. Moreover, different e-Cigarette products need to be tested. The majority of the products tested until now were first-generation atomizers which are prefilled with liquid. This means that the liquid may have been in contact with the metal inside the cartomizer for months, resulting in increased corrosion compared to the tank-type atomizers which are sold empty and are refilled by consumers. Research on the latter products could identify whether new developments have improved the safety profile of e-Cigarettes.

CONCLUSION

Regulation of e-Cigarette chemistry is a very challenging task. The availability of diverse products, the unprecedented ability of consumers to control different functional and performance characteristics of the devices, and the association between perceived satisfaction and efficacy in smoking substitution are important hurdles in setting standardized parameters of use and related aerosol emissions through regulatory decisions. However, there is the potential to apply rules that will dictate the use of high-purity raw ingredients in e-Cigarette liquids. The EU has already implemented the mandatory use of the highest-purity ingredients available for e-Cigarette liquids.[9] Flavorings is an area where research should be intensified because they have a GRAS classification for ingestion but have not, in most cases, been evaluated for inhalation. Compounds with reasonable concern for their safety in inhalation, such as diacetyl and acetyl propionyl, could be restricted as a precautionary measure. Regulation of aerosol emissions seems to be an unfeasible task for now. However, further research and newer technologies could have a valuable regulatory role in reducing toxic emissions and thus establishing additional safety standards. The biggest challenge for regulation will be to identify the fine balance between safety and appeal. A strictly regulated, absolutely safe product with minimal appeal to smokers or low effectiveness as a smoking substitute would have no impact in improving public health. Of course, appeal to nonintended users (i.e., never smokers) should also be considered, and the net population effects should be estimated.

REFERENCES

1. Royal College of Physicians (London) Nicotine without smoke: tobacco harm reduction; 28 April 2016. Available from: https://www.rcplondon.ac.uk/projects/outputs/nicotine-without-smoke-tobacco-harm-reduction-0.

2. Official Journal of the European Communities. DIRECTIVE 2001/37/EC OF THE EUROPEAN PARLIAMENT AND OF THE COUNCIL of 5 June 2001 on the approximation of the laws, regulations and administrative provisions of the Member States concerning the manufacture, presentation and sale of tobacco products. Publication L194/26, 2001. Available from: http://ec.europa.eu/health/tobacco/docs/dir200137ec_tobaccoproducts_en.pdf.

3. Thielen A, Klus H, Muller L. Tobacco smoke: unraveling a controversial subject. *Exp Toxicol Pathol* 2008;**60**:141−56.

4. Borgerding M, Klus H. Analysis of complex mixtures—cigarette smoke. *Exp Toxicol Pathol* 2005;**57**:43−73.

5. European Commission, Scientific Committee on Emerging and Newly Identified Health Risks (SCENIHR). Addictiveness and Attractiveness of Tobacco Additives; 2010. Available from: http://dx.doi.org/10.2772/39398. Available from: http://ec.europa.eu/health/scientific_committees/emerging/docs/scenihr_o_029.pdf.

6. Rabinoff M, Caskey N, Rissling A, Park C. Pharmacological and chemical effects of cigarette additives. *Am J Public Health* 2007;**97**:1981−91.

7. Klein SM, Giovino GA, Barker DC, Tworek C, Cummings KM, O'Connor RJ. Use of flavored cigarettes among older adolescent and adult smokers: United States, 2004−2005. *Nicotine Tob Res* 2008;**10**:1209−14.

8. US Food and Drug Administration (FDA). Candy and fruit flavored cigarettes now illegal in United States; Step is First Under New Tobacco Law; 2009. Available from: http://www.fda.gov/NewsEvents/Newsroom/PressAnnouncements/ucm183211.htm.

9. Official Journal of the European Communities. DIRECTIVE 2014/40/EU OF THE EUROPEAN PARLIAMENT AND OF THE COUNCIL of 3 April 2014 on the approximation of the laws, regulations and administrative provisions of the Member States concerning the manufacture, presentation and sale of tobacco products and repealing Directive 2001/37/EC. Publication L127/1, 2014. Available from: http://ec.europa.eu/health/tobacco/docs/dir_201440_en.pdf.

10. Heck JD, Gaworski CL, Rajendran N, Morrissey RL. Toxicologic evaluation of humectants added to cigarette tobacco: 13-week smoke inhalation study of glycerin and propylene glycol in Fischer 344 rats. *Inhal Toxicol* 2002;**14**:1135−52.

11. Carmines EL, Gaworski CL. Toxicological evaluation of glycerin as a cigarette ingredient. *Food Chem Toxicol* 2005;**43**:1521−39.

12. Gaworski CL, Oldham MJ, Coggins CR. Toxicological considerations on the use of propylene glycol as a humectant in cigarettes. *Toxicology* 2010;**269**:54−66.

13. Renne RA, Yoshimura H, Yoshino K, Lulham G, Minamisawa S, Tribukait A, et al. Effects of flavoring and casing ingredients on the toxicity of mainstream cigarette smoke in rats. *Inhal Toxicol* 2006;**18**:685−706.

14. Vanscheeuwijck PM, Teredesai A, Terpstra PM, Verbeeck J, Kuhl P, Gerstenberg B, et al. Evaluation of the potential effects of ingredients added to cigarettes. Part 4: subchronic inhalation toxicity. *Food Chem Toxicol* 2002;**40**:113–31.

15. Coggins CR, Liu J, Merski JA, Werley MS, Oldham MJ. A comprehensive evaluation of the toxicology of cigarette ingredients: aliphatic and aromatic carboxylic acids. *Inhal Toxicol* 2011;**23**(Suppl. 1):119–40.

16. Carmines EL, Lemus R, Gaworski CL. Toxicologic evaluation of licorice extract as a cigarette ingredient. *Food Chem Toxicol* 2005;**43**:1303–22.

17. Baker RR. The generation of formaldehyde in cigarettes—overview and recent experiments. *Food Chem Toxicol* 2006;**44**:1799–822.

18. Talhout R, Opperhuizen A, van Amsterdam JG. Sugars as tobacco ingredient: effects on mainstream smoke composition. *Food Chem Toxicol* 2006;**44**:1789–98.

19. Baker RR, Massey ED, Smith G. An overview of the effects of tobacco ingredients on smoke chemistry and toxicity. *Food Chem Toxicol* 2004;**42**(Suppl.):S53–83.

20. Talhout R, Opperhuizen A, van Amsterdam JG. Sugars as tobacco ingredient: effects on mainstream smoke composition. *Food Chem Toxicol* 2006;**44**:1789–98.

21. Talhout R, Opperhuizen A, van Amsterdam JG. Role of acetaldehyde in tobacco smoke addiction. *Eur Neuropsychopharmacol* 2007;**17**:627–36.

22. Sershen H, Shearman E, Fallon S, Chakraborty G, Smiley J, Lajtha A. The effects of acetaldehyde on nicotine-induced transmitter levels in young and adult brain areas. *Brain Res Bull* 2009;**79**:458–62.

23. FCTC WHO Framework Concention on Tobacco Control. Regulation of the contents of tobacco products and of tobacco product disclosures. Art. 9/10 guidelines for implementation; 2013. Available from: http://apps.who.int/iris/bitstream/10665/80510/1/9789241505185_eng.pdf.

24. Burns DM, Dybing E, Gray N, Hecht S, Anderson C, Sanner T, et al. Mandated lowering of toxicants in cigarette smoke: a description of the World Health Organization TobReg proposal. *Tob Control* 2008;**17**:132–41.

25. Stratton K, Shetty P, Wallace R, Bondurant S. Clearing the smoke. Assessing the science base for tobacco harm reduction—executive summary. *Tob Control* 2001;**10**:189–95.

26. National Cancer Institute. *Risk associated with smoking cigarettes with low machine-measured yields of tar and nicotine. Smoking and tobacco control monograph No. 13*. Washington, DC: US Department of Health and Human Services, Public Health Service, National Institutes of Health National Cancer Institute; 2001.

27. United States Food and Drug Administration (FDA). Harmful and potentially harmful constituents in tobacco products and tobacco smoke: Established List; April 2012. Available from: http://www.fda.gov/TobaccoProducts/GuidanceComplianceRegulatoryInformation/ucm297786.htm.

28. United States Food and Drug Administration (FDA). Substantial Equivalence. Available from: http://www.fda.gov/TobaccoProducts/Labeling/TobaccoProductReviewEvaluation/SubstantialEquivalence/default.htm [updated April 2016].

29. Benowitz NL, Henningfield JE. Establishing a nicotine threshold for addiction. The implications for tobacco regulation. *N Engl J Med* 1994;**331**:123–5.

30. Henningfield JE, Benowitz NL, Connolly GN, Davis RM, Gray N, Myers ML, et al. Reducing tobacco addiction through tobacco product regulation. *Tob Control* 2004;**13**:132–5.

31. Benowitz NL, Henningfield JE. Reducing the nicotine content to make cigarettes less addictive. *Tob Control* 2013;**22**(Suppl. 1):i14–17.

32. Donny EC, Hatsukami DK, Benowitz NL, Sved AF, Tidey JW, Cassidy RN. Reduced nicotine product standards for combustible tobacco: building an empirical basis for effective regulation. *Prev Med* 2014;**68**:17−22.

33. Benowitz NL, Henningfield JE. Reducing the nicotine content to make cigarettes less addictive. *Tob Control* 2013;**22**(Suppl. 1):i14−7.

34. World Health Organization Study Group of Tobacco Product Regulation. *Advisory note: global nicotine reduction strategy.* Geneva, Switzerland: World Health Organization; 2015. Available from: http://www.who.int/tobacco/publications/prod_regulation/nicotine-reduction/en.

35. Baer E. Fischer HOL. L(+) propylene glycol. *J Am Chem Soc* 1948;**70**:609−10.

36. Lide DR. *CRC handbook of chemistry and physics.* 88th Edition. CRC Press, Talylor & Francis Group; 2007. ISBN 9780849304880.

37. Wurtz A. Mémoire sur les Glycols ou Alcools Diatomiques. *Ann Chim Phys* 1859;**55**:400−78.

38. Organization for Economic Cooperation and Development. Screening Information Data Set for 1,2-Dihydroxypropane (CAS nr 57-55-6); 2001. Available from: http://www.inchem.org/documents/sids/sids/57-55-6.pdf.

39. The Dow Chemical Company. A guide to propylene glycols; 2003. Available from: http://msdssearch.dow.com/PublishedLiteratureDOWCOM/dh_091b/0901b8038091b508.pdf?filepath =propyleneglycol/pdfs/noreg/117-01682.pdf&fromPage=GetDoc.

40. National Institute of Occupational Safety and Health (NIOSH). Health hazard Evaluation Report No. 90-0355-2449; August 1994. Available from: https://www.cdc.gov/niosh/hhe/reports/pdfs/1990-0355-2449.pdf.

41. DHHS/NTP-CERHR. NTP-CERHR Monograph on the Potential Human Reproductive and Developmental Effects of Propylene Glycol; March 2004. NIH Pub No. 04-4482. Available from: https://ntp.niehs.nih.gov/ntp/ohat/egpg/propylene/pg_monograph.pdf.

42. American Medical Association Drug Evaluations. *Prepared by the AMA department of drugs in cooperation with the American Society for Clinical Pharmacology and Therapeutics.* 3rd Ed. Littleton, MA: Publishing Sciences Group; 1977. Available from: http://dx.doi.org/10.1002/jps.2600661054.

43. Wang T, Noonberg S, Steigerwalt R, Lynch M, Kovelesky RA, Rodríguez CA, et al. Preclinical safety evaluation of inhaled cyclosporine in propylene glycol. *J Aerosol Med* 2007;**20**:417−28.

44. Niven R, Lynch M, Moutvic R, Gibbs S, Briscoe C, Raff H. Safety and toxicology of cyclosporine in propylene glycol after 9-month aerosol exposure to beagle dogs. *J Aerosol Med Pulm Drug Deliv* 2011;**24**:205−12.

45. FDA. Generally recognized as safe. 1982; 21 CFR 184.1666. Available from: https://www.gpo.gov/fdsys/pkg/CFR-2000-title21-vol3/pdf/CFR-2000-title21-vol3-sec184-1666.pdf.

46. National Institute of Health. Hazardous Substances Data Bank (HSDB): propylene glycol. Available from: https://toxnet.nlm.nih.gov/cgi-bin/sis/search2/f?./temp/~30iS4f:1.

47. United Nations Joint Expert Committee on Food Additives (JECFA). Evaluation of certain food additives and contaminants: fifty-seventh report of the Joint FAO/WHO expert committee on food additives; 2002. Available from: http://apps.who.int/iris/bitstream/10665/42578/1/WHO_TRS_909.pdf.

48. Branen AL, Davidson PM, Salminen S. *Food additives.* Marcel Dekker Inc; 1990. ISBN-13: 978-0824780463.

49. WHO Technical Report Series No. 539. Toxicological evaluation of certain food with a review of general principles and of specifications. Geneva: World Health Organization; 1974. Available from: http://apps.who.int/iris/bitstream/10665/41072/1/WHO_TRS_539.pdf.

50. American Industrial Hygiene Association (AIHA). *Workplace Environmental Exposure Level (WEEL) Guide for Propylene Glycol*. Fairfax, VA: American Industrial Hygiene Association; 1985.

51. Agency for Toxic Substances and Disease Registry (ATSDR). *Toxicological profile for propylene glycol*. Atlanta, GA; September 1997. Available from: http://www.atsdr.cdc.gov/ToxProfiles/tp189.pdf.

52. Christopher MM, Eckfeldt JH, Eaton JW. Propylene glycol ingestion causes D-lactic acidosis. *Lab Invest* 1990;**62**:114–18.

53. Morshed KM, Nagpaul JP, Majumdar S, Amma MK. Kinetics of oral propylene glycol-induced acute hyperlactatemia. *Biochem Med Metab Biol* 1989;**42**:87–94.

54. Morshed KM, Helgoualch AL, Nagpaul JP, Amma MK, Desjeux JF. The role of propylene glycol metabolism in lactatemia in the rabbit. *Biochem Med Metabol Biol* 1991;**46**:145–51.

55. Ruddick JA. Toxicology, metabolism, and biochemistry of 1,2-propanediol. *Toxicol Appl Pharmacol* 1972;**21**:102–11.

56. LaKind JS, McKenna EA, Hubner RP, Tardiff RG. A review of the comparative mammalian toxicity of ethylene glycol and propylene glycol. *Crit Rev Toxicol* 1999;**29**:331–65.

57. Arbour R, Esparis B. Osmolar gap metabolic acidosis in a 60-year-old man treated for hypoxemic respiratory failure. *Chest* 2000;**118**:545–6.

58. Speth PA, Vree TB, Neilen NF, de Mulder PH, Newell DR, Gore ME, et al. Propylene glycol pharmacokinetics and effects after intravenous infusion in humans. *Ther Drug Monit* 1987;**9**:255–8.

59. Warshaw EM, Botto NC, Maibach HI, Fowler Jr JF, Rietschel RL, Zug KA, et al. Positive patch-test reactions to propylene glycol: a retrospective cross-sectional analysis from the North American Contact Dermatitis Group, 1996 to 2006. *Dermatitis* 2009;**20**:14–20.

60. Lolin Y, Francis DA, Flanagan RJ, Little P, Lascelles PT. Cerebral depression due to propylene glycol in a patient with chronic epilepsy—the value of the plasma osmolal gap in diagnosis. *Postgrad Med J* 1988;**64**:610–13.

61. Lim TY, Poole RL, Pageler NM. Propylene glycol toxicity in children. *J Pediatr Pharmacol Ther* 2014;**19**:277–82.

62. Arulanantham K, Genel M. Central nervous system toxicity associated with ingestion of propylene glycol. *J Pediatr* 1978;**93**:515–16.

63. Zosel A, Egelhoff E, Heard K. Severe lactic acidosis after an iatrogenic propylene glycol overdose. *Pharmacotherapy* 2010;**30**:219.

64. Jorens PG, Demey HE, Schepens PJ, Coucke V, Verpooten GA, Couttenye MM, et al. Unusual D-lactic acid acidosis from propylene glycol metabolism in overdose. *J Toxicol Clin Toxicol* 2004;**42**:163–9.

65. Neale BW, Mesler EL, Young M, Rebuck JA, Weise WJ. Propylene glycol-induced lactic acidosis in a patient with normal renal function: a proposed mechanism and monitoring recommendations. *Ann Pharmacother* 2005;**39**:1732–6.

66. Miller MA, Forni A, Yogaratnam D. Propylene glycol-induced lactic acidosis in a patient receiving continuous infusion pentobarbital. *Ann Pharmacother* 2008;**42**:1502–6.

67. United States Patent Application Publication. Electronic atomization cigarette. Publication No: US 2007/0267031 A1; November 22, 2007.

68. Barbeau AM, Burda J, Siegel M. Perceived efficacy of e-cigarettes versus nicotine replacement therapy among successful e-cigarette users: a qualitative approach. *Addict Sci Clin Pr* 2013;**8**:5.

69. Farsalinos KE, Baeyens F. Harmful effects from one puff of shisha-pen vapor: methodological and interpretational problems in the risk assessment analysis. *Tob Induc Dis* 2016;**14**:22.

70. Etter JF. Throat hit in users of the electronic cigarette: an exploratory study. *Psychol Addict Behav* 2016;**30**:93–100.

71. Li Q, Zhan Y, Wang L, Leischow SJ, Zeng DD. Analysis of symptoms and their potential associations with e-liquids' components: a social media study. *BMC Public Health* 2016;**16**:674.

72. Farsalinos KE, Romagna G, Tsiapras D, Kyrzopoulos S, Voudris V. Characteristics, perceived side effects and benefits of electronic cigarette use: a worldwide survey of more than 19,000 consumers. *Int J Environ Res Public Health* 2014;**11**:4356–73.

73. Robertson OH, Bigg E, Miller BF, Baker Z. Sterilization of air by certain glycols employed as aerosols. *Science* 1941;**93**:213.

74. Henle W, Zellat J. Effect of propylene glycol aerosol on air-borne virus of Influenza. *Proc Soc Exper Biol Med* 1941;**48**:544.

75. Robertson OH, Loosli CG, Puck TT, Bigg E, Miller BF. The protection of mice against Infection with air-borne Influenza virus by means of propylene glycol vapour. *Science* 1941;**94**:612.

76. Harris TH, Stokes Jr. J. The effect of propylene glycol vapour on the incidence of respiratory infections in a convalescent home for children: preliminary observations. *Am J Med Sci* 1942;**204**:430.

77. Harris TH, Stokes Jr. J. Air-borne cross infection in the case of the common cold: a further clinical study of the use of glycol vapours for air sterilization. *Am J Med Sci* 1943;**200**:631.

78. Robertson OH, Bigg E, Puck TT, Miller BF, Technical Assistance of Elizabeth A. Appell. The bactericidal action of propylene glycol vapor on microorganisms suspended in air. I. *J Exp Med* 1942;**75**:593–610.

79. Puck TT, Robertson OH, Lemon HM. The bactericidal action of propylene glycol vapor on microorganisms suspended in air: II. the influence of various factors on the activity of the vapor. *J Exp Med* 1943;**78**:387–406.

80. Puck TT, Wise H, Robertson OH. A device for automatically controlling the concentration of glycol vapors in the air. *J Exp Med* 1944;**80**:377–81.

81. Robertson O.H., Loosli C.G., Puck T.T., Wise H., Lemon H.M., Lester W. Jr. Tests for the chronic toxicity of propylene glycol and triethylene glycol on monkeys and rats by vapor inhalation and oral administration. *J Pharmacol Exp Ther* 1947;**91**: 52–76.

82. Suber RL, Deskin R, Nikiforov I, Fouillet X, Coggins CR. Subchronic nose-only inhalation study of propylene glycol in Sprague-Dawley rats. *Food Chem Toxicol* 1989;**27**:573–83.

83. Wieslander G, Norbäck D, Lindgren T. Experimental exposure to propylene glycol mist in aviation emergency training: acute ocular and respiratory effects. *Occup Environ Med* 2001;**58**:649–55.

84. The Dow Chemical Company. Propylene glycol. Aircraft Deicing Fluid. Available from: http://www.dow.com/propyleneglycol/applications/aircraft_deicing_fluid.htm.

85. Pellegrino R, Viegi G, Brusasco V, Crapo RO, Burgos F, Casaburi R, et al. Interpretative strategies for lung function tests. *Eur Respir J* 2005;**26**:948–68.

86. Lane LB. Freezing points of glycerol and its aqueous solutions. *Ind Eng Chem* 1925;**17**:924.

87. IUPAC-IUB Commission on Biochemical Nomenclature (CBN). Nomenclature of Lipids; 1976. Available from: http://www.chem.qmul.ac.uk/iupac/lipid/.

88. Miller S. *The Soapmaker's companion: a comprehensive guide with recipes, techniques & know-how*. North Adams, MA: Storey Books; 1997. ISBN-13: 978-0882669656.

89. The Soap and Detergent Association (SDA). Glycerine: an overview; 1990. Available from: http://www.aciscience.org/docs/glycerine_-_an_overview.pdf.

90. Wang ZX, Zhuge J, Fang H, Prior BA. Glycerol production by microbial fermentation: a review. *Biotechnol Adv* 2001;**19**:201−23.

91. Christoph R., Schmidt B., Steinberner U., Dilla W., Karinen R. (2006). Glycerol. Ullmann's Encyclopedia of Industrial Chemistry; 2006. ISBN 3527306730.

92. Organization for Economic Cooperation and Development. Screening information data set for glycerol (CAS nr 56-81-5); 2002. Available from: http://www.inchem.org/documents/sids/sids/56815.pdf.

93. Niles D. A glycerin factor. *Biodiesel Mag* 2005. Available from: http://www.biodieselmagazine.com/articles/377/a-glycerin-factor/.

94. Ciriminna R, Della Pina C, Rossi M, Pagliaro M. Understanding the glycerol market. *Eur J Lipid Sci Technol* 2014;**116**:1432−9.

95. Hedinger. Specification: Glycerol Ph. Eur./USP/JP. Available from: http://www.hedinger.de/fileadmin/user_upload/downloads/Produkte/Produktspezifikation/083_Glycerol_EP_USP_JP_parenteral_grade_040414.pdf.

96. Lin ECC. Glycerol utilization and its regulation in mammals. *Ann Rev Biochem* 1977;**46**:765−95.

97. Tao RC, Kelley RE, Yoshimura NN, Benjamin F. Glycerol: its metabolism and use as intravenous energy source. *J Parent Enter Nutr* 1983;**7**:479−88.

98. Bortz WM, Paul P, Haff AC, Holmes WL. Glycerol turnover and oxidation in man. *J Clin Invest* 1972;**51**:1537−46.

99. The Cohen Group. Recommended exposure guidelines for glycol fogging agents. Project No. 6070-1001; 1997. Available from: http://www.lemaitreltd.com/pdf/Cohen%20Report.pdf.

100. Hine CH, Anderson HH, Moon HD, Dunlap MK, Morse MS. Comparative toxicity of synthetic and natural glycerin. *AMA Arch Ind Hyg Occup Med* 1953;**7**:282−91.

101. Clark CR, Marshall TC, Merickel BS, Sanchez A, Brownstein DG, Hobbs CH. Toxicological assessment of heat transfer fluids proposed for use in solar energy applications. *Toxicol Appl Pharmacol* 1979;**51**:529−35.

102. Meyer JS, Charney JZ, Rivera VM, Mathew NT. Treatment with glycerol of cerebral oedema due to acute cerebral infarction. *Lancet* 1971;**2**:993−7.

103. MacDonald I. Effects of dietary glycerol on the serum glyceride level of men and women. *Br J Nutr* 1970;**24**:537−43.

104. Renne RA, Wehner AP, Greenspan BJ, Deford HS, Ragan HA, Westerberg RB, et al. 2-Week and 13-week inhalation studies of aerosolized glycerol in rats. *Inhal Toxicol* 1992;**4**:95−111.

105. Dawkins L, Turner J, Roberts A, Soar K. "Vaping" profiles and preferences: an online survey of electronic cigarette users. *Addiction* 2013;**108**:1115−25.

106. Farsalinos KE, Romagna G, Tsiapras D, Kyrzopoulos S, Spyrou A, Voudris V. Impact of flavour variability on electronic cigarette use experience: an internet survey. *Int J Environ Res Public Health* 2013;**10**:7272−82.

107. The Flavor and Extract Manufacturers Association (FEMA). About the FEMA GRAS Program. Available from: http://www.femaflavor.org/gras.

108. Smith RL, Adams TB, Cohen SM, Doull J, Feron VJ, Goodman JI, et al. Safety evaluation of natural flavor complexes. *Toxicol Lett* 2004;**149**:197−207.

109. Smith RL, Cohen SM, Doull J, Feron VJ, Goodman JI, Marnett LJ, et al. Criteria for the safety evaluation of flavoring substances. The Expert Panel of the Flavor and Extract Manufacturers Association. *Food Chem Toxicol* 2005;**43**:1141−77.

110. US Food and Drug Administration (FDA). Overview of food ingredients, additives & colors; November 2004 (revised April 2010). Available from: http://www.fda.gov/Food/IngredientsPackagingLabeling/FoodAdditivesIngredients/ucm094211.htm.

111. European Union. Regulation (EC) No. 2232/96 of the European Parliament and of the Council of 28 October 1996 laying down a Community procedure for flavouring substances used or intended for use in or on foodstuffs. *Official J Eur Comm L299* 1996. Available from: http://eur-lex.europa.eu/legal-content/EN/TXT/PDF/?uri=CELEX:31996R2232.

112. EFSA Panel on Food Contact Materials, Enzymes, Flavourings and Processing Aids. Guidance on the data required for the risk assessment of flavourings to be used in or on foods. *EFSA J* 2010;**8**:1623. Available from: http://onlinelibrary.wiley.com/doi/10.2903/j.efsa.2010.1623/epdf.

113. Zhu SH, Sun JY, Bonnevie E, Cummins SE, Gamst A, Yin L, et al. Four hundred and sixty brands of e-cigarettes and counting: implications for product regulation. *Tob Control* 2014;**23**(Suppl. 3):iii3–9.

114. The Flavor and Extract Manufacturers Association (FEMA). Safety assessment and regulatory authority to use flavors: focus on e-cigarettes; 2016. Available from: http://www.femaflavor.org/safety-assessment-and-regulatory-authority-use-flavors-focus-e-cigarettes.

115. The Flavor and Extract Manufacturers Association (FEMA). Respiratory health and safety in the flavor manufacturing workplace; 2012 Update. Available from: http://www.femaflavor.org/sites/default/files/linked_files/FEMA_2012%20Respiratory%20Health%20and%20Safety.pdf.

116. Farsalinos KE, Kistler KA, Gillman G, Voudris V. Why we consider the NIOSH-proposed safety limits for diacetyl and acetyl propionyl appropriate in the risk assessment of electronic cigarette liquid use: a response to Hubbs et al. *Nicotine Tob Res* 2015;**17**: 1290–11291.

117. National Institute for Occupational Safety and Health (NIOSH). (2011). Criteria for a recommended standard: occupational exposure to diacetyl and 2,3-pentanedione. Available from: http://www.cdc.gov/niosh/docket/archive/pdfs/NIOSH-245/DraftDiacetylCriteriaDocument081211.pdf.

118. Diaz A, Ventura F, Galceran MT. Identification of 2,3-butanedione (diacetyl) as the compound causing odor events at trace levels in the Llobregat River and Barcelona's treated water (Spain). *J Chromatogr A* 2004;1034:175–182.

119. Blank I, Sen A, Grosch W. Potent odorants of the roasted powder and brew of arabica coffee. *Z Lebensm Unters Forsch* 1992;**195**:239–45.

120. Mathews JM, Watson SL, Snyder RW, Burgess JP, Morgan DL. Reaction of the butter flavorant diacetyl (2-3 butanedione) with N-α-Acetylarginine: a model for epitope formation of pulmonary protein in the etiology of obliterative bronchiolitis. *J Agric Food Chem* 2010;**58**:12761–8.

121. National Toxicology Program. Chemical information review document for artificial butter flavoring and constituents: diacetyl [CAS No. 431-03-8] and acetoin [CAS No. 513-86-0]; 2007. Available from: http://ntp.niehs.nih.gov/ntp/htdocs/chem_background/exsumpdf/artificial_butter_flavoring.pdf.

122. Nakagawa J, Ishikura S, Asami J, Isaji T, Usami N, Hara A, et al. Molecular characterization of mammalian dicarbonyl/L-xylulose reductase and its localization in 45 kidney. *J Biol Chem* 2002;**277**:17883–91.

123. Otsuka M, Mine T, Ohuchi K, Ohmori S. A detoxication route for acetaldehyde: metabolism of diacetyl, acetoin, and 2,3-butanediol in liver homogenate and perfused liver of rats. *J Biochem (Tokyo)* 1996;**119**:246–51.

124. Sawada H, Hara A, Nakayama T, Seiriki K. Kinetic and structural properties of diacetyl reductase from hamster liver. *J Biochem* 1985;**98**:1349–57.

125. Morris JB, Hubbs AF. Inhalation dosimetry of diacetyl and butyric acid, two components of butter flavoring vapors. *Toxicol Sci* 2009;**108**:173–83.

126. Gloede E, Cichocki JA, Baldino JB, Morris JB. A validated hybrid computational fluid dynamics-physiologically based pharmacokinetic model for respiratory tract vapor absorption in the human and rat and its application to inhalation dosimetry of diacetyl. *Toxicol Sci* 2011;**123**:231–46.

127. Endo S, Matsunaga T, Horie K, Tajima K, Bunai Y, Carbone V, et al. Enzymatic characteristics of an aldo-keto reductase family protein (AKR1C15) and its localization in rat tissues. *Arch Biochem Biophys* 2007;**465**:136–47.

128. Weigt SS, DerHovanessian A, Wallace WD, Lynch 3rd JP, Belperio JA. Bronchiolitis obliterans syndrome: the Achilles' heel of lung transplantation. *Semin Respir Crit Care Med* 2013;**34**:336–51.

129. Schlesinger C, Meyer CA, Veeraraghavan S, Koss MN. Constrictive (obliterative) bronchiolitis: diagnosis, etiology, and a critical review of the literature. *Ann Diagn Pathol* 1998;**2**:321–34.

130. Estenne M, Maurer JR, Boehler A, Egan JJ, Frost A, Hertz M, et al. Bronchiolitis obliterans syndrome 2001: an update of the diagnostic criteria. *J Heart Lung Transplant* 2002;**21**:297–310.

131. Kreiss K, Gomaa A, Kullman G, Fedan K, Simoes EJ, Enright PL. Clinical bronchiolitis obliterans in workers at a microwave-popcorn plant. *N Engl J Med* 2002;**23**:347, 330–338.

132. Castro-Rodriguez JA, Giubergia V, Fischer GB, Castaños C, Sarria EE, Gonzalez R, et al. Postinfectious bronchiolitis obliterans in children: the South American contribution. *Acta Paediatr* 2014;**103**:913–21.

133. Yu J. Postinfectious bronchiolitis obliterans in children: lessons from bronchiolitis obliterans after lung transplantation and hematopoietic stem cell transplantation. *Korean J Pediatr* 2015;**58**:459–65.

134. Saber H, Saburi A, Ghanei M. Clinical and paraclinical guidelines for management of sulfur mustard induced bronchiolitis obliterans; from bench to bedside. *Inhal Toxicol* 2012;**24**:900–6.

135. National Institute of Occupational Safety and Health (NIOSH). *Hazard evaluation and technical assistance report*. South Bend, IN: International Bakers Services, Inc; 1985. Available from: https://www.cdc.gov/niosh/hhe/reports/pdfs/1985-0171-1710.pdf.

136. Centers for Disease Control and Prevention (CDC). Fixed obstructive lung disease in workers at a microwave popcorn factory—Missouri, 2000–2002. *MMWR Morb Mortal Wkly Rep* 2002;**51**:345–7.

137. Parmet AJ. Bronchiolitis in popcorn-factory workers. *N Engl J Med* 2002;**347**:1980–2.

138. Ezrailson EG. Bronchiolitis in popcorn-factory workers. *N Engl J Med* 2002;**347**:1980–2 author reply 1980–1982; discussion 1980–1982.

139. Akpinar-Elci M, Travis W1D, Lynch DA, Kreiss K. Bronchiolitis obliterans syndrome in popcorn production plant workers. *Eur Respir J* 2004;**24**:298–302.

140. Kanwal R. Bronchiolitis obliterans in workers exposed to flavoring chemicals. *Curr Opin Pulm Med* 2008;**14**:141–6.

141. Kanwal R, Kullman G, Fedan K, Kreiss K. Occupational lung disease risk and exposures to butter flavoring chemicals after implementation of controls at a microwave popcorn plant. *Public Health Rep* 2011;**126**:480–94.

142. van Rooy FG, Smit LA, Houba R, Zaat VA, Rooyackers JM, Heederik DJ. A cross-sectional study of lung function and respiratory symptoms among chemical workers producing diacetyl for food flavourings. *Occup Environ Med* 2009;**66**:105–10.

143. Halldin CN, Suarthana E, Fedan KB, Lo YC, Turabelidze G, Kreiss K. Increased respiratory disease mortality at a microwave popcorn production facility with worker risk of bronchiolitis obliterans. *PLoS One* 2013;**8**:e57935.

144. Hubbs AF, Battelli LA, Goldsmith WT, Porter DW, Frazer D, Friend S, et al. Necrosis of nasal and airway epithelium in rats inhaling vapors of artificial butter flavoring. *Toxicol Appl Pharmacol* 2002;**185**:128−35.

145. Hubbs AF, Goldsmith WT, Kashon ML, Frazer D, Mercer RR, Battelli LA, et al. Respiratory toxicologic pathology of inhaled diacetyl in Sprague-Dawley rats. *Toxicol Pathol* 2008;**36**:330−44.

146. Morgan DL, Flake GP, Kirby PJ, Palmer SM. Respiratory toxicity of diacetyl in 26 C57BL/6 mice. *Toxicol Sci* 2008;**103**:169−80.

147. Palmer SM, Flake GP, Kelly FL, Zhang HL, Nugent JL, Kirby PJ, et al. Severe airway epithelial injury, aberrant repair and bronchiolitis obliterans develops after diacetyl instillation in rats. *PLoS One* 2011;**6**:e17644.

148. Day G, LeBouf R, Grote A, Pendergrass S, Cummings K, Kreiss K, et al. Identification and measurement of diacetyl substitutes in dry bakery mix production. *J Occup Environ Hyg* 2011;**8**:93−103.

149. Zaccone EJ, Thompson JA, Ponnoth DS, Cumpston AM, Goldsmith WT, Jackson MC, et al. Popcorn flavoring effects on reactivity of rat airways in vivo and in vitro. *J Toxicol Environ Health A* 2013;**76**:669−89.

150. Morgan DL, Jokinen MP, Price HC, Gwinn WM, Palmer SM, Flake GP. Bronchial and bronchiolar fibrosis in rats exposed to 2,3-pentanedione vapors: implications for bronchiolitis obliterans in humans. *Toxicol Pathol* 2012;**40**:448−65.

151. Morgan DL, Jokinen MP, Johnson CL, Price HC, Gwinn WM, Bousquet RW, et al. Chemical reactivity and respiratory toxicity of the α-diketone flavoring agents: 2,3-butanedione, 2,3-pentanedione, and 2,3-hexanedione. *Toxicol Pathol* 2016;**44**:763−83.

152. Hubbs AF, Cumpston AM, Goldsmith WT, Battelli LA, Kashon ML, Jackson MC, et al. Respiratory and olfactory cytotoxicity of inhaled 2,3-pentanedione in Sprague-Dawley rats. *Am J Pathol* 2012;**181**:829−44.

153. European Commission. Recommendation from the Scientific Committee on Occupational Exposure Limits for diacetyl; June 2014. Available from: http://ec.europa.eu/social/BlobServlet?docId=6511.

154. Gaffney SH, Abelman A, Pierce JS, Glynn ME, Henshaw JL, McCarthy LA, et al. Naturally occurring diacetyl and 2,3-pentanedione concentrations associated with roasting and grinding unflavored coffee beans in a commercial setting. *Toxicol Rep* 2015;**2**:1171−81.

155. Fujioka K, Shibamoto T. Determination of toxic carbonyl compounds in cigarette smoke. *Environ Toxicol* 2006;**21**:47−54.

156. Moree-Testa P, Saint-Jalm Y. Determination of [alpha]-dicarbonyl compounds in cigarette smoke. *J Chromatogr* 1981;**217**:197−208.

157. Pierce JS, Abelmann A, Spicer LJ, Adams RE, Finley BL. Diacetyl and 2,3-pentanedione exposures associated with cigarette smoking: implications for risk assessment of food and flavoring workers. *Crit Rev Toxicol* 2014;**44**:420−35.

158. Farsalinos KE, Kistler KA, Gillman G, Voudris V. Evaluation of electronic cigarette liquids and aerosol for the presence of selected inhalation toxins. *Nicotine Tob Res* 2015;**17**:168−74.

159. Hubbs AF, Cummings KJ, McKernan LT, Dankovic DA, Park RM, Kreiss K. Comment on Farsalinos et al., "Evaluation of Electronic Cigarette Liquids and Aerosol for the Presence of Selected Inhalation Toxins". *Nicotine Tob Res* 2015;**17**:1288−9.

160. Allen JG, Flanigan SS, LeBlanc M, Vallarino J, MacNaughton P, Stewart JH, et al. Flavoring chemicals in e-cigarettes: diacetyl, 2,3-pentanedione, and acetoin in a sample of 51 products, including fruit-, candy-, and cocktail-flavored e-cigarettes. *Environ Health Perspect* 2016;**124**:733−9.

161. Kanwal R, Kullman G, Piacitelli C, Sahakian N, Martin S, Fedan K, et al. Evaluation of flavorings-related lung disease risk at six microwave popcorn plants. *J Environ Med* 2006;**48**:149−57.

162. Raherison C, Girodet PO. Epidemiology of COPD. *Eur Respir Rev* 2009;**18**:213−21.

163. Niewoehner DE, Kleinerman J, Rice DB. Pathologic changes in the peripheral airways of young cigarette smokers. *N Engl J Med* 1974;**291**:755−8.

164. Farsalinos KE, Yannovits N, Sarri T, Voudris V, Poulas K. Protocol proposal for, and evaluation of, consistency in nicotine delivery from the liquid to the aerosol of electronic cigarettes atomizers: regulatory implications. *Addiction* 2016;**111**:1069−76.

165. Talih S, Balhas Z, Eissenberg T, Salman R, Karaoghlanian N, El Hellani A, et al. Effects of user puff topography, device voltage, and liquid nicotine concentration on electronic cigarette nicotine yield: measurements and model predictions. *Nicotine Tob Res* 2015;**17**:150−7.

166. Lopez AA, Hiler MM, Soule EK, Ramôa CP, Karaoghlanian NV, Lipato T, et al. Effects of electronic cigarette liquid nicotine concentration on plasma nicotine and puff topography in tobacco cigarette smokers: a preliminary report. *Nicotine Tob Res* 2016;**18**:720−3.

167. Shihadeh A, Eissenberg T. Electronic cigarette effectiveness and abuse liability: predicting and regulating nicotine flux. *Nicotine Tob Res* 2015;**17**:158−62.

168. Balster RL, Schuster CR. Fixed-interval schedule of cocaine reinforcement: effect of dose and infusion duration. *J Exp Anal Behav* 1973;**20**:119−29.

169. Carter LP, Stitzer ML, Henningfield JE, O'Connor RJ, Cummings KM, Hatsukami DK. Abuse liability assessment of tobacco products including potential reduced exposure products. *Cancer Epidemiol Biomarkers Prev* 2009;**18**:3241−62.

170. Farsalinos KE, Voudris V, Le Houezec J. Risks of attempting to regulate nicotine flux in electronic cigarettes. *Nicotine Tob Res* 2015;**17**:163−4.

171. Dawkins LE, Kimber CF, Doig M, Feyerabend C, Corcoran O. Self-titration by experienced e-cigarette users: blood nicotine delivery and subjective effects. *Psychopharmacology (Berl)* 2016;**233**:2933−41.

172. European Directorate for the Quality of Medicines and Healthcare. European Pharmacopeia 7.0; 2012.

173. United Stated Pharmacopeia Monographs: Nicotine. U.S. Pharmacopeia. Available from: www.pharmacopeia.cn/v29240/usp29nf24s0_m56620.html.

174. Etter JF, Zäther E, Svensson S. Analysis of refill liquids for electronic cigarettes. *Addiction* 2013;**108**:1671−9.

175. Lisko JG, Tran H, Stanfill SB, Blount BC, Watson CH. Chemical composition and evaluation of nicotine, tobacco alkaloids, pH, and selected flavors in e-cigarette cartridges and refill solutions. *Nicotine Tob Res* 2015;**17**:1270−8.

176. Kisaki T, Maeda S, Koiwai A, Mikami Y, Sasaki T, Matsushita H. Transformation of tobacco alkaloids. *Beitr Tabakforsch Int* 1978;**9**:308−16.

177. Linnell RH. The oxidation of nicotine. I. Kinetics of the liquid phase reaction near room temperature. *Tob Sci* 1960;**4**:89−91.

178. Fowler JS, Volkow ND, Wang G-J, Pappas N, Logan J, Shea C, et al. Brain monoamine oxidase A inhibition in cigarette smokers. *Proc Natl Acad Sci USA* 1996;**93**:14065−9.

179. Guillem K, Vouillac C, Azar MR, Parsons LH, Koob GF, Cador M, et al. Monoamine oxidase inhibition dramatically increases the motivation to self-administer nicotine in rats. *J Neurosci* 2005;**25**:8593−600.

180. Rose JE, Behm FM, Ramsey C, Ritchie Jr. JC. Platelet monoamine oxidase, smoking cessation and tobacco withdrawal symptoms. *Nicotine Tob Res* 2001;**3**:383−90.

181. Belluzzi JD, Wang R, Leslie FM. Acetaldehyde enhances acquisition of nicotine self-administration in adolescent rats. *Neuropsychopharmacology* 2005;**30**:705−12.

182. Talhout R, Opperhuizen A, van Amsterdam JG. Role of acetaldehyde in tobacco smoke addiction. *Eur Neuropsychopharmacol* 2007;**17**:627−36.

183. Herraiz T, Chaparro C. Human monoamine oxidase is inhibited by tobacco smoke: beta-carboline alkaloids act as potent and reversible inhibitors. *Biochem Biophys Res Commun* 2005;**326**:378−86.

184. Herraiz T. Relative exposure to beta-carbolines norharman and harman from foods and tobacco smoke. *Food Addit Contam* 2004;**21**:1041−50.

185. Clemens KJ, Caillé S, Stinus L, Cador M. The addition of five minor tobacco alkaloids increases nicotine-induced hyperactivity, sensitization and intravenous self-administration in rats. *Int J Neuropsychopharmacol* 2009;**12**:1355−66.

186. Smith TT, Schaff MB, Rupprecht LE, Schassburger RL, Buffalari DM, Murphy SE, et al. Effects of MAO inhibition and a combination of minor alkaloids, β-carbolines, and acetaldehyde on nicotine self-administration in adult male rats. *Drug Alcohol Depend* 2015;**155**:243−52.

187. Kramlinger VM, von Weymarn LB, Murphy SE. Inhibition and inactivation of cytochrome P450 2A6 and cytochrome P450 2A13 by menthofuran, betanicotyrine and menthol. *Chemicobiol Interact* 2012;**197**:87−92.

188. Su T, Bao Z, Zhang QY, Smith TJ, Hong JY, Ding X. Human cytochrome P450 CYP2A13: predominant expression in the respiratory tract and its high efficiency metabolic activation of a tobacco-specific carcinogen, 4-(methylnitrosamino)-1-(3-pyridyl)-1-butanone. *Cancer Res* 2000;**60**:5074−9.

189. Di YM, Chow VD, Yang LP, Zhou SF. Structure, function, regulation and polymorphism of human cytochrome P450 2A6. *Curr Drug Metab* 2009;**10**:754−80.

190. Abramovitz A, McQueen A, Martinez RE, Williams BJ, Sumner W. Electronic cigarettes: the nicotyrine hypothesis. *Med Hypotheses* 2015;**85**:305−10.

191. Uchiyama S, Ohta K, Inaba Y, Kunugita N. Determination of carbonyl compounds generated from the e-cigarette using coupled silica cartridges impregnated with hydroquinone and 2,4-dinitrophenylhydrazine, followed by high-performance liquid chromatography. *Analyt Sci* 2013;**29**:1219−22.

192. Lauterbach JH, Spencer A. Generation of acetaldehyde and other carbonyl compounds during vaporization of glycerol and propylene glycol during puffing of a popular style of e-cigarette. Society of Toxicology National Meeting, San Diego, CA; 2015.

193. Kosmider L, Sobczak A, Fik M, Knysak J, Zaciera M, Kurek J, et al. Carbonyl compounds in electronic cigarette vapors: effects of nicotine solvent and battery output voltage. *Nicotine Tob Res* 2014;**16**:1319−26.

194. Farsalinos KE, Voudris V, Poulas K. E-cigarettes generate high levels of aldehydes only in 'dry puff' conditions. *Addiction* 2015;**110**:1352−6.

195. Geiss O, Bianchi I, Barrero-Moreno J. Correlation of volatile carbonyl yields emitted by e-cigarettes with the temperature of the heating coil and the perceived sensorial quality of the generated vapours. *Int J Hyg Environ Health* 2016;**219**:268−77.

196. Jensen RP, Luo W, Pankow JF, Strongin RM, Peyton DH. Hidden formaldehyde in e-cigarette aerosols. *N Engl J Med* 2015;**372**:392−4.

197. Romagna G, Allifranchini E, Bocchietto E, Todeschi S, Esposito M, Farsalinos KE. Cytotoxicity evaluation of electronic cigarette vapor extract on cultured mammalian fibroblasts (ClearStream-LIFE): comparison with tobacco cigarette smoke extract. *Inhal Toxicol* 2013;**25**:354−61.

198. Farsalinos KE, Romagna G, Tsiapras D, Kyrzopoulos S, Voudris V. Evaluation of electronic cigarette use (vaping) topography and estimation of liquid consumption: implications for research protocol standards definition and for public health authorities' regulation. *Int J Environ Res Public Health* 2013;**10**:2500−14.

199. Bates CD, Farsalinos KE. E-cigarettes need to be tested for safety under realistic conditions. *Addiction* 2015;**110**:1688−9.

200. United States of America, Department of Health and Human Services, Food and Drug Administration, Center for Tobacco Products. Electronic cigarettes and the Public Health: a public workshop; March 9, 2015. Farsalinos K. E-cigarette use topography: variation among users in realistic conditions; changes according to power levels; association with nicotine absorption. Transcript 3.9.15, pages 160−171. Available from: http://www.fda.gov/downloads/TobaccoProducts/NewsEvents/UCM442808.pdf.

201. Georgia State University, Department of Physics and Astronomy. Hyperphysics. Resistance: Temperature Coefficient. Available from: http://hyperphysics.phy-astr.gsu.edu/hbase/electric/restmp.html.

202. Steam engine. Available from: http://www.steam-engine.org/wirewiz.asp.

203. Goniewicz ML, Knysak J, Gawron M, Kosmider L, Sobczak A, Kurek J, et al. Levels of selected carcinogens and toxicants in vapour from electronic cigarettes. *Tob Control* 2014;**23**:133−9.

204. Williams M, Villarreal A, Bozhilov K, Lin S, Talbot P. Metal and silicate particles including nanoparticles are present in electronic cigarette cartomizer fluid and aerosol. *PLoS One* 2013;**8**:e57987.

205. Farsalinos KE, Voudris V, Poulas K. Are metals emitted from electronic cigarettes a reason for health concern? A risk-assessment analysis of currently available literature. *Int J Environ Res Public Health* 2015;**12**:5215−32.

206. Williams M, To A, Bozhilov K, Talbot P. Strategies to reduce tin and other metals in electronic cigarette aerosol. *PLoS One* 2015;**10**:e0138933.

207. Medicines and Healthcare products Regulatory Agency (UK). E-cigarette working group discussion paper on submission of notifications under article 20 of directive 2014/40/EU. Chapter 3: emissions from electronic cigarettes. Available from: https://www.gov.uk/government/uploads/system/uploads/attachment_data/file/544094/3_Emissions_disc_paper_final.pdf.

CHAPTER 6

Potential Impact of e-Cigarette Usage on Human Health

R. Polosa

INTRODUCTION

Cigarette smoke contains more than 7000 chemicals, many of which are known to be harmful to the human body, thus causing a broad range of fatal diseases.[1] Death is mainly from ischemic heart disease, stroke, lung cancer, and the catastrophic complications of advanced-stage chronic obstructive pulmonary disease (COPD).[1,2]

Abstention from smoking is known to lower the risk of developing these diseases and to produce significant health gains in patients who already have them.[1,2] Most smokers want to quit and many make attempts to do so, but most of these attempts fail largely because of the powerful addictive qualities of nicotine and nonnicotine sensory and behavioral cues.[3,4] For those willing to quit, a combination of pharmacotherapy and behavioral intervention can support attempts and double or triple the quit rate.[5,6] However, outside the context of rigorous randomized controlled trials (RCTs), efficacy rates are disappointingly low, with an estimated annual population cessation rate of approximately 4−5%.[7−9]

e-Cigarettes (ECs) are becoming an increasingly attractive long-term alternative to conventional cigarettes because of their many similarities with smoking behavior.[10,11] Recent internet-based surveys[12,13] and clinical trials[14,15] show that ECs may help smokers reduce or abstain from cigarette consumption.

Although vapor toxicology is by far less problematic than that of conventional cigarettes (reviewed in[16]) and e-vapor products are at least 95% less harmful than combustible cigarettes,[17] there is concern as to whether chronic exposure to their residual toxicological load may

Analytical Assessment of e-Cigarettes. DOI: http://dx.doi.org/10.1016/B978-0-12-811241-0.00006-1

nevertheless carry a risk for human health. The safety of long-term EC use is a legitimate clinical question.

Only large prospective clinical studies of well-characterized EC users will determine the long-term health impact of these products. But such studies are quite demanding due to several methodological, logistical, ethical, and financial challenges. In particular, to address the potential of future disease development, hundreds of users would need to be followed for many years before any conclusions could be made. We cannot wait 50 years before getting measurements of the effects of decades of vaping. Moreover, characterization of EC users in these studies should take into account detailed information on their previous exposure to tobacco smoking (adult vapers are almost invariably ex-smokers). Indeed, it would be very difficult if not impossible to establish whether reported changes in regular EC users are related to chronic exposure to e-vapor or to their previous smoking history without a study population made up of regular vapers who have never smoked. Alternatively, some stratification for the pack-years (this parameter gives an estimate of the total amount a person has been exposed to cigarette smoke and provides an idea of the overall risk related to tobacco use) in the vapers (i.e., ex-smokers) is mandatory to dissect out responses driven by chronic exposure to e-vapor from those related to previous smoking history. Others strategies must be pursued.

A much less challenging approach is to explore responses to e-vapor in human cell lines in vitro and in animal models. However, these approaches also have inherent flaws. Findings cannot be applied directly to humans because of the inability to test the normal consumption exposure conditions of e-vapor products, the fact that standards for vapor production and exposure protocols have not been clearly defined, and the risk of over- or underestimation in the interpretation of toxic effects. Consequently, it is not surprising to find a divergence in the literature, with some authors reporting little or no injury[18] and others describing much higher levels of toxicity and inflammatory responses despite using the same cell lines.[19]

Great care must be taken in projecting results from animal studies to humans, especially if the animals are very dissimilar (e.g., rodents rather than primates). As with cell studies, the comparison between different things being tested is probably of greater value than speculative projection of a finding from animal to human. For example, in a recent study, mice exposed to EC aerosol emissions had greater susceptibility to airway infection.[20] The impaired bacterial and viral response was more likely to be the consequence of protracted stress and nicotine overdosing in the EC-exposed animals. Most importantly, these interesting findings in rodents do not corroborate the evidence in humans. Despite millions of regular EC users, there is no conclusive published evidence of greater susceptibility to airway infection. Quite the opposite, in fact: A recent online survey showed that switching from smoking to vaping is likely to be associated with a reduced incidence of self-reported respiratory infections.[21] Last but not least, animal models do not take into account any potential impact that prior smoking history might have had on the measured responses. This is important because harm that has accumulated throughout many years of smoking does not disappear at the point of switching to vaping or quitting completely and may well introduce bias and lead to erroneous interpretation.

As previously mentioned, it would take hundreds of well-characterized EC users to be followed prospectively for many years to investigate the long-term health effects of ECs. Given that this is highly impractical, alternatively it is possible either to detect early changes of subclinical injury in "healthy" smokers switching to EC with highly sensitive functional tests, or to explore changes in EC users who have preexisting disease with more robust and validated investigational tools. In this chapter, an overview of the emerging health findings from people who have been switched from conventional cigarettes to ECs in experimental studies and in clinical trials will be presented. The evidence will be subjected to critical appraisal focusing on methodological issues, safety concerns, and potential benefits deriving from the regular use of ECs. The literature search included peer-reviewed literature from the National Center for Biotechnology Information's PubMed website and official reports and monographs from reputable governmental organizations that the author has recently appraised as part of a major on-going interest in ECs, risk reduction, and harm reversal. Keywords used in the search were smoking cessation and reduction, e-Cigarette, spirometry, respiratory symptoms, exhaled biomarkers, asthma, COPD, allergy, blood pressure, arterial stiffness, hypertension, body weight, post-cessation weight gain, nicotine, tobacco harm reduction, and harm reversal.

ACUTE STUDIES—RESPIRATORY SYSTEM

Before discussing acute studies, it must be emphasized that transient effects on health measurements are generally not very informative, as they cannot establish causation of clinically relevant harm nor have prognostic value.

Findings from Internet surveys and clinical trials have reported transient throat irritation, dry cough, and other symptoms of respiratory irritation in some smokers when switching to ECs (reviewed in[16]). Although it is possible that unknown contaminants or byproducts contained in the EC aerosol may cause nonspecific irritant effects, these are likely to occur in EC users who might have developed propylene glycol hypersensitivity.[22]

Hence the prompt defensive response against irritants from EC aerosol is the most likely cause for the immediate physiologic changes detected with highly sensitive respiratory functional tests as reported by Vardavas et al.[23] The question of whether such an acute irritation could translate into clinically meaningful lung disease remains unanswered, but there certainly is no evidence to suggest that this is an indication of clinically significant adverse lung effects. For example, it is highly questionable that a trivial (well within test variability) 16% decrease in exhaled nitric oxide levels (2.1 ppb in absolute terms) and 11% increase in peripheral flow resistance (0.025 kPa/L/second in absolute terms) from baseline after using an EC for 5 minutes will have clinical relevance.[24,25] More importantly, no significant changes have been detected by standard validated respiratory function parameters (i.e., forced expiratory volume in the first second (FEV1), forced vital capacity (FVC), FEV% (FEV1/FVC index), and peak expiratory flow (PEF)) after EC use.[23] Lack of a significant acute effect on airflow obstruction by these four measures after short-term EC use has been confirmed recently.[26,27] Although the results on the transient effects of exhaled nitric oxide have been generally conflicting,[23,26,28,29] switching to EC use quickly and universally leads to a near-normalization in toxic levels of exhaled carbon monoxide (reviewed in[16]).

Overall, studies that have focused on the acute effect of ECs on lung function and airway responses do not appear to support negative respiratory health outcomes in EC users.

ACUTE STUDIES—CARDIOVASCULAR SYSTEM

Given the important causal link between cigarette smoking and cardiovascular disease,[30–32] it is not surprising that researchers have concentrated on acute changes in cardiovascular vital signs and indices of vascular function after EC use. Changes to heart rate or blood pressure (BP) after short periods of EC use have been investigated in several acute studies,[33–41] but no consistent changes have been reported overall due to differences in exposure protocols, e-liquid nicotine concentrations, and hardware design.

Whereas a randomized crossover study of 32 smokers who trialed four products (an EC with a 1.8% nicotine cartridge, an EC with a 1.6% nicotine cartridge, chosen brand of conventional cigarette, and an unlit cigarette)[33] failed to show any statistically significant increase in heart rate after 10 puffs of EC, a more recent randomized crossover trail of 23 smokers who were asked to try six products (five ECs containing from 1.6% to 2.4% nicotine and varying levels of propylene glycol and glycerine, and a conventional cigarette)[34] showed significant increases in heart rate after 50 puffs of 2.4% nicotine EC (but not after 1.6% nicotine EC). In the same study, there was a significant increase in diastolic BP from baseline after use of the EC products (mean increase ranging from 3.17 to 6.83 mmHg), but surprisingly no significant increase was reported for systolic BP.[34] Lack of significant changes in heart rate, systolic BP, or diastolic BP has been confirmed in two small nonrandomized crossover studies immediately after ad lib EC use for 5 minutes[37] or 15 puffs for 10 minutes.[38] A nonrandomized trial of 40 regular EC users (all former smokers) using an EC containing 11 mg nicotine/mL e-liquid for 7 minutes ad lib[39] found no significant changes to mean heart rate (67.1–67.5 bpm) or systolic BP (123.9–124.6 mmHg) from baseline after EC use. Surprisingly, a small but significant change in mean diastolic BP was observed (from 75.6 to 78.5 mmHg). In the same study, the authors also examined the immediate effects of EC use on several echocardiographic parameters of diastolic function.[39] Diastolic function is a very sensitive measure of subclinical cardiac dysfunction. As expected, cigarette smoking caused diastolic dysfunction, but none of the parameters investigated were changed by vaping.

On the other hand, a crossover single-blind study of 40 "healthy" subjects (20 smokers, 20 nonsmokers, matched for age and sex) showed that 9 puffs of EC can increase markers of oxidative stress and elicit vascular dysfunction (measured by flow-mediated dilatation).[40] The effects observed after EC use were slightly less pronounced than those caused by conventional tobacco cigarettes of similar nominal nicotine content. The observed acute changes in vascular function after tobacco smoking and EC use are likely to be attributed to nicotine. Nonetheless, it is worth pointing out that effects of similar magnitude are also observed after coffee consumption.[41] A comparison with coffee would have made a worthwhile additional arm to this study and may have helped with interpreting whether there is a material risk of harm.

More recently, acute changes in vascular function after EC use were studied by assessing aortic stiffness (investigated by carotid-femoral pulse-wave velocity) in 24 "healthy" smokers.[42] As expected, pulse-wave velocity increased after a 5-minutes ad lib session of cigarette smoking (by 0.44 m/second); this was more pronounced than the change measured after EC use (by 0.19 m/second). When ad lib EC use was protracted for a period of 30 minutes (please note that vaping continuously for 30 minutes does not reflect realistic use of these products), acceleration in pulse-wave velocity became similar to that measured after cigarette smoking (by 0.36 m/second). Once again, the reported findings on aortic stiffness after tobacco smoking and EC use are due to the acute effects of nicotine and comparable to those occurring after drinking coffee.[43]

The effects described in these latter studies do not constitute cardiovascular harm, as they result from a transient vascular response to nicotine, and certainly do not call into question the emerging evidence base that vaping is much less harmful to the cardiovascular system than smoking.

LONG-TERM STUDIES—RESPIRATORY SYSTEM

Long-term changes in lung function, airway responses, and respiratory symptoms have been monitored in a 1-year prospective RCT of "healthy" smokers who were invited to quit or reduce their tobacco consumption by switching to ECs.[44,45] No evidence of airways obstruction at any of the subsequent follow-up visits could be observed, irrespective of participants' smoking phenotype classification (see Fig. 6.1). This is not unexpected, given that study participants were "healthy" smokers without preexisting lung disease. Nonetheless, early

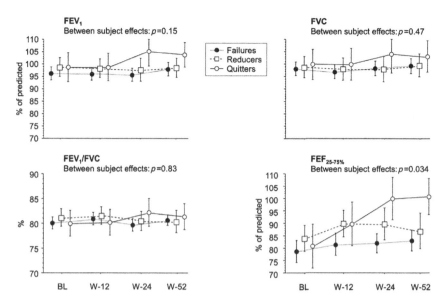

Figure 6.1 Means (±95% confidence intervals) of FEV1, FVC, FEV1/FVC, and FEF25–75% at baseline (BL) and weeks 12 (W-12), 24 (W-24), and 52 (W-52), separately for smoking phenotype (failures: closed circles; reducers: open squares; quitters: open circles).

improvements in FEF25–75% (a sensitive measure of obstruction in the more peripheral airways) were detected at 3 months after switching among those who completely gave up cigarette smoking, with steady progressive improvements being observed at 6 and 12 months (lower right panel).[44] Specifically, improvements in FEF25–75% were no different in quitters who stopped using ECs compared with quitters who were still using ECs. Interestingly, the progressive normalization of peripheral airways function was associated with a substantial reduction in self-reported respiratory symptoms (cough and shortness of breath), particularly in those who completely gave up smoking.[44] It is not known whether harm reversal in peripheral airways can translate into efficient prevention of airway disease in the future.

Low levels of nitric oxide[46,47] and high levels of carbon monoxide[48] are generally found in the exhaled breath of cigarette smokers, with levels returning to the normal nonsmoking range after quitting. In a recent 1-year RCT of "healthy" smokers who were invited to quit or reduce their cigarette consumption by switching to ECs, normalization of both exhaled nitric oxide (see Fig. 6.2) and carbon monoxide levels (see Fig. 6.3) were noted among those who completely gave up

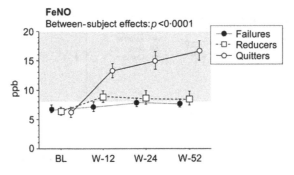

Figure 6.2 Time trends (means ±95% confidence intervals) of exhaled nitric oxide (FeNO) at baseline (BL) and follow-up visits at weeks 12 (W-12), 24 (W-24), and 52 (W-52), separately for smoking phenotype (failures, reducers, and quitters). The smoking phenotype at W-52 (between-subjects factor) showed a significant effect by repeated-measures ANOVA. Shaded area delineates FeNO reference ranges for "healthy" nonsmokers.

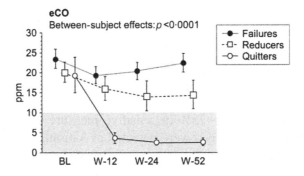

Figure 6.3 Time trends (means ±95% confidence intervals) of exhaled carbon monoxide (eCO) at baseline (BL) and follow-up visits at weeks 12 (W-12), 24 (W-24), and 52 (W-52), separately for smoking phenotype (failures, reducers, and quitters). The smoking phenotype at W-52 (between-subjects factor) showed a significant effect by repeated-measures ANOVA. Shaded area delineates commonly accepted eCO reference ranges for "healthy" nonsmokers.

cigarette smoking.[45] Reversal to within normal nonsmoking levels was documented at 3 months and completed at 6 and 12 months.[45] Similar normalization in exhaled nitric oxide and carbon monoxide levels was observed in quitters who stopped using ECs as well as quitters who were still using ECs. On the other hand, no significant changes were observed in failures and reducers. That completely abstaining from cigarette smoking leads to a remarkable fall in toxic levels of exhaled carbon monoxide to within normal limits is in agreement with earlier observations in acute[33,49] and long-term studies.[15,50] Given that ECs are battery-operated devices that work without burning tobacco, this was not surprising. Of note, the reported improvements in exhaled nitric oxide and carbon monoxide levels were correlated with attenuations in composite symptom scores (symptoms included cough, phlegm, shortness of breath, wheeze, tight chest, stuffy nose, sinus pain, and frontal headache), particularly in those who completely gave up smoking. The mechanism for the observed improvements in symptom scores is likely to be related to the reversal of inflammatory changes in the upper and lower airways when giving up smoking. Hence considering that exhaled nitric oxide and carbon monoxide levels are known to be elevated during airway inflammation,[51–53] it was not surprising to observe a significant relationship with symptom score improvements. Given that nitric oxide is known to inhibit replication of respiratory pathogens[54] and that low nitric oxide levels might explain the high incidence in lower respiratory tract infection in those exposed to active or passive smoking,[28,55] that increases in the exhaled carbon monoxide levels may also reflect airway inflammation,[53] and that emerging data have revealed several new mechanisms for carbon monoxide and nitric oxide in tumor biology,[56] the observed normalization in smokers who were invited to quit or reduce tobacco consumption by switching to ECs might contribute to a reduction in the risk of lower respiratory tract infection and lung cancer. However, future studies will be required to show whether normalization of these exhaled biomarkers translates into efficient prevention of respiratory tract infections and significant reduction in lung cancer risk.

Only limited data are available on the health effects of EC use in users with preexisting pulmonary diseases.

The asthmatic smoker is a distinct disease phenotype with increased susceptibility to exacerbations and poor asthma-specific health status.[57] Increased disease severity and marked impairment in asthma control is more frequently reported in asthmatic smokers who have smoked more than 20 pack-years.[58] Most studies show an accelerated decline in lung function and increased airflow obstruction,[59] and asthma patients who smoke appear to have an impaired response to the beneficial effects of antiasthma drugs compared to asthmatics who do not.[60,61] Quitting smoking can reverse the negative impact of tobacco smoke on asthma symptoms and lung function,[62] and switching to ECs use may produce significant respiratory benefits.

Recent retrospective clinical studies conducted to ascertain the efficacy and safety of regular EC use in mild to moderate asthma failed to detect deterioration in respiratory physiology and subjective asthma outcomes.[63,64] On the contrary, smokers with asthma who quit or substantially decreased tobacco consumption by switching to regular ECs showed progressive significant improvement in the Juniper's Asthma Control Questionnaire (ACQ), FEV1, FVC, forced expiratory flow measured between 25% and 75% of forced vital capacity (FEF25–75%), as well as airway hyperresponsiveness (AHR) to inhaled methacholine throughout the 1-year reporting period.[63] A 2-year follow-up confirmed that EC use ameliorates objective and subjective asthma outcomes and shows that these beneficial effects may persist in the long term (Fig. 6.4A–E).[64]

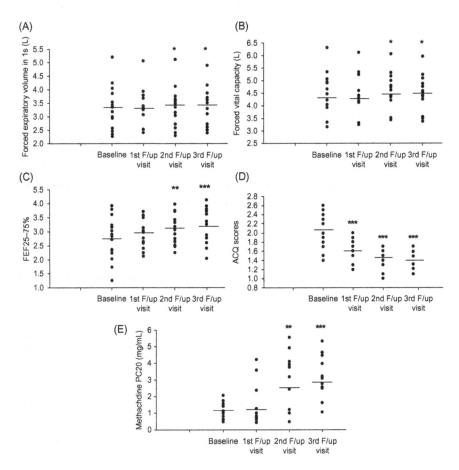

Figure 6.4 Dot plots of individual data points in a study of 16 ED users with asthma. (A) Forced expiratory volume (FEV1); (B) Forced vital capacity (FVC); (C) Forced expiratory flow at the middle half of the FVC (FEF25–75%); (D) Juniper's Asthma Control Questionnaire (ACQ) scores; (E) Methacholine PC20 response. The lines in the panels show the means. *$p \leq 0.05$; **$p \leq 0.01$; ***$p \leq 0.001$ compared to baseline. F/up, follow-up; L/s, liters per second.

Interestingly, similar findings were found in the dual users. EC use was very well tolerated, and exposure to e-vapor in this vulnerable population did not trigger any asthma attacks.

The positive findings are consistent with results from a large internet survey of EC users diagnosed with asthma.[12] An improvement in symptoms of asthma after switching was reported in 65.4% of the respondents. Although improved asthma symptoms were more often noted in exclusive EC users, similar improvements were also described in dual users. Worsening after switching was reported only in 1.1% of the asthmatics.

Taken together, these findings provide emerging evidence that EC use can reverse the harmful effects of tobacco smoking on the asthmatic airways. It is plausible that the attenuation in pro-inflammatory effects of cigarette smoke on the airways by switching to EC use might have caused the observed improvement in lung function and asthma symptoms.

COPD is a progressive disease characterized by a persistent inflammatory and remodeling response of the airways that is generally associated with tobacco smoking,[65,66] with about 15–20% of smokers developing full blown COPD.[67] Smoking cessation is the only evidence-based strategy known to improve COPD prognosis.[67,68] Smoking cessation reduces the rate of annual decline in pulmonary function, attenuates respiratory symptoms of cough and sputum, and improves health status.[69–71] Switching to ECs use may yield significant respiratory benefits in COPD.

Formal efficacy and safety assessment of EC use was recently conducted in a retrospective clinical study of patients with COPD. No deterioration in respiratory physiology (postbronchodilator FEV1, FVC, and %FEV1/FVC) was observed in COPD patients who quit or reduced substantially their tobacco consumption by switching to vaping (R Polosa, personal observation). The lack of significant changes in standard spirometric indices after smoking cessation is not unusual in smokers with COPD.[72,73] Nonetheless, progressive significant improvement in COPD exacerbation, annual decline in FEV1, overall health status (measured by CAT), and physical activity (measured by the 6-minute walk distance test) was documented throughout the 2-year reporting period.

The reported improvement in health outcomes is in agreement with findings from an internet survey of 1062 regular EC users diagnosed with COPD.[12] Improvement in respiratory symptoms after switching was reported in 75.7% of the respondents, whereas worsening was reported in only 0.8%.

That respiratory exacerbations were halved in COPD patients who quit or reduced substantially their tobacco consumption by adopting vaping is a key informative finding. Chronic exposure of the airways to cigarette smoke is known to promote susceptibility to infection through a number of different mechanisms.[74–76] Thus abstaining from tobacco smoking by switching to ECs may explain the reduction in respiratory infections.[77] Besides, regular vaping may have additional theoretical health benefits, given that propylene glycol in its aerosol form is a potent bactericidal agent. Last but not least, in a recent 1-year prospective RCT, we have shown that it is possible to obtain steady progressive normalization of antimicrobial and antiinflammatory activity in the exhaled breath of smokers who switch to ECs.[45] This may also translate into an efficient prevention of respiratory tract infections and significant reduction of COPD exacerbation.

Although the beneficial effects were consistently documented for each and every health outcome, the results of these small uncontrolled studies in self-selected patients with asthma and COPD must be interpreted with caution. Only long-term evaluation of objective and subjective respiratory health outcomes in a large sample of well-characterized patients will provide a definitive answer about the long-term impact on lung health. But the current evidence generally supports a beneficial effect of EC use in respiratory patients.

LONG-TERM STUDIES—CARDIOVASCULAR SYSTEM

The interaction between smoking and BP is complex and there is controversy over the independent chronic effect of smoking on BP.[78,79] In already established hypertension, smoking is associated with a heightened risk for cardiovascular disease; thus quitting smoking is unquestionably among the most important steps patients with elevated BP can take to improve their cardiovascular health.[80–82] Surprisingly, however, data on the long-term effects of smoking cessation or reduction on BP (and HR) is very limited, and results are unclear, with studies reporting lower, higher, or unchanged BP values in smokers compared with nonsmokers.[83] In particular, population studies have important methodological limitations that may predispose to heterogeneous results because (1) they rely on self-reported tobacco use and casual collection of BP measurements, (2) the observed relationship between levels of smoking and changes in BP in a cross-sectional design does not imply causation, and (3) such studies do not take into account common confounders (e.g., age, gender, weight increase, caffeine, and alcohol intake) that may play a crucial role when determining potential causation.

Long-term changes in BP and heart rate were monitored in a recent 1-year prospective RCT of "healthy" smokers who were invited to quit or reduce their tobacco consumption by switching to ECs.[84] A small reduction in systolic BP from baseline was observed at week 52 in the whole study population ($n = 145$), irrespective of smoking phenotype classification (quitters, reducers, failures). This is not surprising, because detection of improvements in smokers with no history of hypertension and with a normal BP at baseline is highly unlikely. When the same analysis was repeated in the subgroup of 66 subjects who had elevated BP at baseline, a substantial reduction in systolic BP was observed at week 52 (132.4 vs 141.2 mmHg), with a significant effect now being found for smoking phenotype classification (Fig. 6.5). After adjusting for weight change, gender, and age, reduction in systolic BP remained significantly associated with both smoking reduction

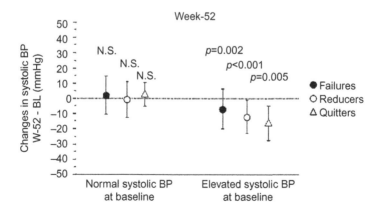

Figure 6.5 Change (mean ±SD, absolute mmHg) in systolic blood pressure (SBP) from baseline to week 52 for continuous smoking phenotypes, separately for subjects with normal and elevated SBP at baseline (BL). p-values for significance of changes from baseline are shown.

and smoking abstinence. Of note, there was no difference in BP reduction from baseline between quitters who stopped using ECs and quitters who were still using ECs. The findings are important, since it is well-established that high-normal BP is a risk factor for future development of hypertension, and is associated with an increased risk of myocardial infarction and coronary artery disease.[85,86]

The demonstration that smokers who reduce or quit smoking by using ECs may lower their systolic BP suggests that the harmful effects of cigarette smoke on the vascular system can be reversed. By substantially reducing exposure to conventional cigarettes' hazardous toxicants and achieving clinically relevant BP reductions, EC use may not only improve the cardiovascular risk profile but also confer an overall health advantage in smokers unable or unwilling to quit who are also at risk of developing arterial hypertension, compared to continuing smoking.

Only limited data are available on the health effects of ECs in users with preexisting cardiovascular diseases.

Cigarette smoking is responsible for about 50% of all avoidable deaths in smokers, half of these due to cardiovascular disease.[30] The risk is primarily related to the amount of tobacco smoked daily and shows a clear dose–response relationship with no obvious lower limit for deleterious effects.[31,87] By sustaining low-grade systemic inflammation[88] and contributing to arterial stiffness,[89] tobacco smoke is also likely to lead to arterial hypertension,[90] thus further worsening smokers' cardiovascular risk profile.

Despite the strong relationship between smoking and elevated cardiovascular risk, very little is known about the long-term effects of smoking cessation on BP in already established hypertension. Specifically, it is unknown if regular EC use could result in improved BP control in patients with a diagnosis of hypertension.

A recent retrospective clinical study investigated long-term changes in resting BP and in level of BP control in smokers with a diagnosis of hypertension who quit or reduced substantially their tobacco consumption by switching to ECs (R Polosa, personal observation). As a result of the substantial reduction in cigarette consumption, decreased systolic and diastolic BP as well as improved BP control were reported. In the EC group, systolic and diastolic BP fell by 10 mmHg and

6 mmHg, respectively. Predictably, not only was there a decrease in the proportion of patients with hypertension, but the overall proportion of patients in the EC group with good BP control increased five-fold compared to the reference group. Of note, improvement in resting BP as well as in level of BP control was unrelated to the minor adjustments in the use of antihypertensives, and EC use was well tolerated with no reported severe adverse reactions or acute decompensation in BP. In agreement with the findings from other research groups,[91,92] positive improvements in BP were noted not only in quitters but also in those who reduced consumption in conventional cigarettes (dual users). A possible explanation is that the dual users in our study substantially reduced their average number of cigarettes per day, with 77.3% reporting a reduction of at least 75% from baseline by the end of the observation period. The observed improvement in systolic and diastolic BP and BP control in hypertensive smokers who switched to regular ECs use suggests that the harmful effects of cigarette smoke on the vascular system can be reversed.

Caution is necessary when interpreting the findings of this small uncontrolled study in self-selected patients with arterial hypertension, but the observed improvements were consistently documented throughout the study. Prospective evaluation of objective and subjective outcomes in large samples of well-characterized hypertensive patients will clarify the long-term impact of ECs on cardiovascular health.

LONG-TERM STUDIES—BODY WEIGHT

In view of the reported discrepancy about the long-term effects of smoking cessation on BP,[83,93] improvement in BP after switching to regular EC use requires additional explanation. Besides the predictable methodological limitations of population studies that may predispose to heterogeneous results, it is also important to consider the effect of changes in body weight, a common confounder that plays a crucial role when analyzing BP datasets. Indeed, the association of elevated risk for future development of hypertension after smoking cessation has been mainly attributed to postcessation weight gain.[90,94] Low BP in smokers is related to decreased body weight,[95] with higher body weight and elevated BP being more common in former smokers than nonsmokers.[96]

In the small uncontrolled study in self-selected patients with arterial hypertension discussed previously, postcessation weight gain after switching to regular EC use was so insignificant that it might have contributed to the positive long-term effects of smoking cessation on BP and BP control. Likewise, in the 1-year prospective RCT of "healthy" smokers who were invited to quit or reduce their tobacco consumption by switching to ECs,[84] the observed reduction in systolic BP remains significantly associated with both smoking reduction and smoking abstinence even after weight change, in the multiple linear regression analysis. Given the trivial weight gain in quitters using ECs at week 52 (only about 0.6 kg), this was not surprising.

What is surprising, however, is that although smoking cessation is known to be associated with significant weight gain,[97–99] it is substantially limited in quitters who use ECs. This hypothesis was recently tested in a 1-year prospective evaluation of body weight change in "healthy" smokers who were invited to quit or reduce their tobacco consumption by switching to ECs.[100] Quitters gained on average 2.4, 2.9, and 1.5 kg at weeks 12, 24, and 52, respectively (Fig. 6.6). Thus weight gain after switching was relatively small and much lower than that reported in the literature.[97,101] In a metaanalysis of 62 prospective studies recording weight changes in abstinent

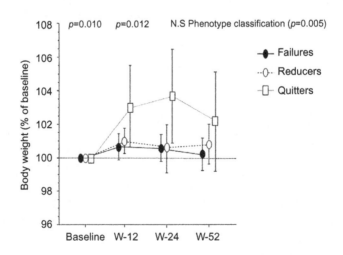

Figure 6.6 Time trends in body weight changes (expressed as percent change from baseline ±95% confidence intervals) at weeks 12 (W-12), 24 (W-24), and s52 (W-52), separately for smoking phenotype (Failures: closed circles, Reducers: open squares; Quitters: open circles). Repeated-measures ANOVA showed that within-factor (body weight) differences were significant (p = 0.005); the effect of between phenotype classification showed a significant effect at weeks 12 (p = 0.010) and 24 (p = 0.012) but not at week 52.

smokers, average weight gain at 3, 6, and 12 months after quitting were 2.9, 4.2, and 4.7 kg, respectively.[101]

The relatively small weight gain could be due to the low incidence of reported increase in appetite in the study.[100] Increased appetite was reported only in 20% of quitters, with a significant effect for postcessation weight gain only at week 12. It is possible that by providing a coping mechanism for conditioned smoking cues, EC use could mitigate hunger associated with smoking abstinence. Moreover, substitution of conventional cigarettes with ECs might limit postcessation weight gain because these products—by mimicking the physical, visual, sensorial, and behavioral experience of conventional smoking—can replace the oral satisfaction without which many abstainers would fill their sense of emptiness and hunger with compulsive eating. Taken together, these mechanisms suggest how EC use might have been the cause for the observed weight gain mitigation in quitters.

Postcessation weight gain can have important health consequences, considering that smoking and obesity are both well-known risk factors for cardiovascular disease and some cancers.[102] Smoking and obesity are also risk factors for type II diabetes.[103,104] Consequently, by substantially reducing tobacco consumption and minimizing postcessation weight gain, EC-based interventions may improve smokers' overall cardiovascular and metabolic risk profile.[105] It is important that research continue in this area because the negative effects of obesity could outweigh the health benefits achieved through reductions in smoking prevalence.[106,107]

REFERENCES

1. US Department of Health and Human Services. *The Health Consequences of Smoking: 50 Years of Progress: A Report of the Surgeon General*. Atlanta, GA: US Department of Health and Human Services, Centers for Disease Control and Prevention, National Center for Chronic Disease Prevention and Health Promotion, Office on Smoking and Health; 2014.

2. Doll R, Peto R, Boreham J, Sutherland I. Mortality in relation to smoking: 50 year observations on male British doctors. *BMJ* 2004;**328**:1519e1528.

3. Buchhalter AR, Acosta MC, Evans SE, Breland AB, Eissenberg T. Tobacco abstinence symptom suppression: the role played by the smoking-related stimuli that are delivered by denicotinized cigarettes. *Addiction* 2005;**100**:550e559.

4. Hughes JR, Keely J, Naud S. Shape of the relapse curve and long-term abstinence among untreated smokers. *Addiction* 2004;**99**:29e38.

5. Polosa R, Benowitz NL. Treatment of nicotine addiction: present therapeutic options and pipeline developments. *Trends Pharmacol Sci* 2011;**32**:281e289.

6. Stead LF, Lancaster T. Combined pharmacotherapy and behavioural interventions for smoking cessation. *Cochrane Database Syst Rev* 2012;**10**:CD008286.

7. Alpert HR, Connolly GN, Biener L. A prospective cohort study challenging the effectiveness of population-based medical intervention for smoking cessation. *Tob Control* 2013;**22**:32e37.

8. Pierce JP, Cummins SE, White MM, Humphrey A, Messer K. Quitlines and nicotine replacement for smoking cessation: do we need to change policy? *Annu Rev Public Health* 2012;**33**:341e356.

9. Zhu SH, Lee M, Zhuang YL, Gamst A, Wolfson T. Interventions to increase smoking cessation at the population level: how much progress has been made in the last two decades? *Tob Control* 2012;**21**:110e118.

10. Caponnetto P, Russo C, Bruno CM, Alamo A, Amaradio MD, Polosa R. Electronic cigarette: a possible substitute for cigarette dependence. *Monaldi Arch Chest Dis* 2013;**79**:12–19 PMID: 23741941.

11. Caponnetto P, Campagna D, Papale G, Russo C, Polosa R. The emerging phenomenon of electronic cigarettes. *Expert Rev Respir Med* 2012;**6**(1):63–74.

12. Farsalinos KE, Romagna G, Tsiapras D, Kyrzopoulos S, Voudris V. Characteristics, perceived side effects and benefits of electronic cigarette use: a worldwide survey of more than 19,000 consumers. *Int J Environ Res Public Health* 2014;**11**:4356–73. PMID: 24758891. Available from: http://dx.doi.org/10.3390/ijerph110404356.

13. Siegel MB, Tanwar KL, Wood KS. Electronic cigarettes as a smoking-cessation: tool results from an online survey. *Am J Prev Med* 2011;**40**:472–5. PMID:21406283. Available from: http://dx.doi.org/10.1016/j.amepre.2010.12.006.

14. Caponnetto P, Campagna D, Cibella F, Morjaria JB, Caruso M, Russo C, et al. EffiCiency and Safety of an eLectronic cigAreTte (ECLAT) as tobacco cigarettes substitute: a prospective 12-month randomized control design study. *PLoS One* 2013;**8**:e66317. PMID:23826093. Available from: http://dx.doi.org/10.1371/journal.pone.0066317.

15. Polosa R, Caponnetto P, Maglia M, Morjaria JB, Russo C. Success rates with nicotine personal vaporizers: a prospective 6-month pilot study of smokers not intending to quit. *BMC Public Health* 2014;**14**:1159. PMID:25380748. Available from: http://dx.doi.org/10.1186/1471-2458-14-1159.

16. Farsalinos KE, Polosa R. Safety evaluation and risk assessment of electronic cigarettes as tobacco cigarette substitutes: a systematic review. *Ther Adv Drug Saf* 2014;**5**:67–86.

17. Nutt DJ, Phillips LD, Balfour D, Curran HV, Dockrell M, Foulds J, et al. Estimating the harms of nicotine-containing products using the MCDA approach. *Eur Addict Res* 2014;**20**:218–25. PMID:24714502. Available from: http://dx.doi.org/10.1159/000360220.

18. Misra M, Leverette RD, Cooper BT, Bennett MB, Brown SE. Comparative in vitro toxicity profile of electronic and tobacco cigarettes, smokeless tobacco and nicotine replacement therapy products: e-liquids, extracts and collected aerosols. *Int J Environ Res Public Health* 2014;**11**:11325–47.

19. Cervellati F, Muresan XM, Sticozzi C, Gambari R, Montanger G, Forman HJ, et al. Comparative effects between electronic and cigarette smoke in human keratinocytes and epithelial lung cells. *Toxicol In Vitro* 2014;**28**:999–1005.

20. Sussan TE, Gajghate S, Thimmulappa RK, Ma J, Kim JH, Sudini K, et al. Exposure to electronic cigarettes impairs pulmonary anti-bacterial and anti-viral defenses in a mouse model. *PLoS One* 2015;**10**(2):e0116861.

21. Miler JA, Mayer B, Hajek P. Changes in the frequency of airway infections in smokers who switched to vaping: results of an online survey. *J Addict Res Ther* 2016;**7**:4. Available from: http://dx.doi.org/10.4172/2155-6105.1000290.

22. Polosa R, Campagna D, Tashkin D. Subacute bronchial toxicity induced by an electronic cigarette: take home message. *Thorax* 2014;**69**:588.

23. Vardavas C, Anagnostopoulos N, Kougias M, Evangelopoulou V, Connolly G, Behrakis P. Short-term pulmonary effects of using an electronic cigarette: impact on respiratory flow resistance, impedance, and exhaled nitric oxide. *Chest* 2012;**141**:1400–6.

24. Oostveen E, MacLeod D, Lorino H, Farré R, Hantos Z, Desager K, et al. The forced oscillation technique in clinical practice: methodology, recommendations and future developments. *Eur Respir J* 2003;**22**:1026–41.

25. Dweik RA, Boggs PB, Erzurum SC, Irvin CG, Leigh MW, Lundberg JO, et al. An official ATS clinical practice guideline: interpretation of exhaled nitric oxide levels (FENO) for clinical applications. *Am J Respir Crit Care Med* 2011;**184**:602–15.

26. Flouris A, Chorti M, Poulianiti K, Jamurtas A, Kostikas K, Tzatzarakis M, et al. Acute impact of active and passive electronic cigarette smoking on serum cotinine and lung function. *Inhal Toxicol* 2013;**25**:91–101.

27. Ferrari M, Zanasi A, Nardi E, Morselli Labate AM, Ceriana P, et al. Short-term effects of a nicotine-free e-cigarette compared to a traditional cigarette in smokers and non-smokers. *BMC Pulm Med* 2015;**15**:120.

28. Marini S, Buonanno G, Stabile L, Ficco G. Short-term effects of electronic and tobacco cigarettes on exhaled nitric oxide. *Toxicol Appl Pharmacol* 2014;**278**:9–15.

29. Schober W, Szendrei K, Matzen W, Osiander-Fuchs H, Heitmann D, Schettgen T, et al. Use of electronic cigarettes (e-cigarettes) impairs indoor air quality and increases FeNO levels of e-cigarette consumers. *Int J Hyg Environ Health* 2014;**217**:628–37.

30. Perk J, De Backer G, Gohlke H, Graham I, Reiner Z, Verschuren M, et al. European guidelines on cardiovascular disease prevention in clinical practice (version 2012). The fifth joint task force of the european society of cardiology and other societies on cardiovascular disease prevention in clinical practice (constituted by representatives of nine societies and by invited experts). *Eur Heart J* 2012;**33**:1635–701.

31. Teo KK, Ounpuu S, Hawken S, Pandey MR, Valentin V, Hunt D, et al. Tobacco use and risk of myocardial infarction in 52 countries in the interheart study: a case-control study. *Lancet* 2006;**368**:647–58.

32. Cacciola RR, Guarino F, Polosa R. Relevance of endothelial-haemostatic dysfunction in cigarette smoking. *Curr Med Chem* 2007;**14**(17):1887–92.

33. Vansickel AR, Cobb CO, Weaver MF, Eissenberg TE. A clinical laboratory model for evaluating the acute effects of electronic "cigarettes": nicotine delivery profile and cardiovascular and subjec- tive effects. *Cancer Epidemiol Biomarkers Prev* 2010;**19**:1945–53.

34. Yan XS, D'Ruiz C. Effects of using electronic cigarettes on nicotine delivery and cardiovascular function in comparison with regular cigarettes. *Regul Toxicol Pharmacol* 2014;**71**:24–34.

35. Vansickel AR, Weaver MF, Eissenberg T. Clinical laboratory assessment of the abuse liability of an electronic cigarette. *Addiction* 2012;**107**:1493–500.

36. Vansickel AR, Eissenberg T. Electronic cigarettes: effective nicotine delivery after acute administration. *Nicotine Tob Res* 2013;**15**:267–70.

37. Czogala J, Cholewinski M, Kutek A, Zielinska-Danch W. Evaluation of changes in hemodynamic parameters after the use of electronic nicotine delivery systems among regular cigarette smokers. *Przegl Lek* 2012;**69**:841–5.

38. Szoltysek-Boldys I, Sobczak A, Zielinska-Danch W, Barton A, Koszowski B, Kosmider L. Influence of inhaled nicotine source on arterial stiffness. *Przegl Lek* 2014;**71**:572–5.

39. Farsalinos KE, Tsiapras D, Kyrzopoulos S, Savvopoulou M, Voudris V. Acute effects of using an electronic nicotine-delivery device (electronic cigarette) on myocardial function: comparison with the effects of regular cigarettes. *BMC Cardiovasc Disord* 2014;**14**:78.

40. Carnevale R, Sciarretta S, Violi F, Nocella C, Loffredo L, Perri L, et al. Acute impact of tobacco vs electronic cigarette smoking on oxidative stress and vascular function. *Chest* 2016;**150**(3):606−12.

41. Papamichael CM, Aznaouridis KA, Karatzis EN, Karatzi KN, Stamatelopoulos KS, Vamvakou G, et al. Effect of coffee on endothelial function in healthy subjects: the role of caffeine. *Clin Sci (Lond)* 2005;**109**(1):55−60.

42. Vlachopoulos C, Ioakeimidis N, Abdelrasoul M, Terentes-Printzios D, Georgakopoulos C, Pietri P, et al. Electronic cigarette smoking increases aortic stiffness and blood pressure in young smokers. *J Am Coll Cardiol* 2016;**67**(23):2802−3.

43. Mahmud A, Feely J. Acute effect of caffeine on arterial stiffness and aortic pressure waveform. *Hypertension* 2001;**38**(2):227−31.

44. Cibella F, Campagna D, Caponnetto P, Amaradio MD, Caruso M, Russo C, et al. Lung function and respiratory symptoms in a randomized smoking cessation trial of electronic cigarettes. *Clin Sci (Lond)* 2016. pii: CS20160268. [Epub ahead of print] PubMed PMID: 27543458.

45. Campagna D, Cibella F, Caponnetto P, Amaradio MD, Caruso M, Morjaria JB, et al. Changes in breathomics from a 1-year randomized smoking cessation trial of electronic cigarettes. *Eur J Clin Invest* 2016;**46**(8):698−706. Available from: http://dx.doi.org/10.1111/eci.12651. Epub 2016 Jul 8. PubMed PMID: 27322745.

46. Robbins RA, Millatmal T, Lassi K, Rennard S, Daughton D. Smoking cessation is associated with an increase in exhaled nitric oxide. *Chest* 1997;**112**:313−18.

47. Travers J, Marsh S, Aldington S, Williams M, Shirtcliffe P, Pritchard A, et al. Reference ranges for exhaled nitric oxide derived from a random community survey of adults. *Am J Respir Crit Care Med* 2007;**176**:238−42.

48. Jarvis MJ, Russell MA, Saloojee Y. Expired air carbon monoxide: a simple breath test of tobacco smoke intake. *BMJ* 1980;**281**:484−5.

49. McRobbie H, Phillips A, Goniewicz ML, Myers Smith K, Knight-West O, Przulj D, et al. Effects of switching to electronic cigarettes with and without concurrent smoking on exposure to nicotine, carbon monoxide, and acrolein. *Cancer Prev Res* 2015;**8**:873−8.

50. Polosa R, Morjaria J, Caponnetto P, Campagna D, Russo C, Alamo A, et al. Effectiveness and tolerability of electronic cigarette in real-life: a 24-month prospective observational study. *Intern Emerg Med* 2014;**9**:537−46.

51. Alving K, Weitzberg E, Lundberg JM. Increased amount of nitric oxide in exhaled air of asthmatics. *Eur Respir J* 1993;**6**:1368−70.

52. Kharitonov SA, Yates D, Robbins RA, Logan-Sinclair R, Shinebourne EA, Barnes PJ. Increased nitric oxide in exhaled air of asthmatic patients. *Lancet* 1994;**343**:133−5.

53. Zhang J, Yao X, Yu R, Bai J, Sun Y, Huang M, et al. Exhaled carbon monoxide in asthmatics: a meta-analysis. *Respir Res* 2010;**11**:50.

54. Nathan CF, Hibbs JBJ. Role of nitric oxide synthesis in macrophage antimicrobial activity. *Curr Opin Immunol* 1991;**3**:65−70.

55. Shiva F, Nasiri M, Sadeghi B, Padyab M. Effects of passive smoking on common respiratory symptoms in young children. *Acta Paediatr* 2003;**92**:1394−7.

56. Szabo C. Gasotransmitters in cancer: from pathophysiology to experimental therapy. *Nat Rev Drug Discov* 2016;**15**:185−203.

57. Polosa R, Thomson NC. Smoking and asthma: dangerous liaisons. *Eur Respir J* 2013;**41**:716−26.

58. Polosa R, Russo C, Caponnetto P, Bertino G, Sarva M, Antic T, et al. Greater severity of new onset asthma in allergic subjects who smoke: a 10-year longitudinal study. *Respir Res* 2011;**12**. Available from: http://dx.doi.org/10.1186/1465-9921-12-16.

59. Lange P, Parner J, Vestbo J, Schnohr P, Jensen G. A 15-year follow-up study of ventilatory function in adults with asthma. *N Engl J Med* 1998;**339**:1194–200.

60. Tomlinson JE, McMahon AD, Chaudhuri R, Thompson JM, Wood SF, Thomson NC. Efficacy of low and high dose inhaled corticosteroid in smokers versus non-smokers with mild asthma. *Thorax* 2005;**60**:282–7.

61. Chaudhuri R, Livingston E, McMahon AD, Thomson L, Borland W, Thomson NC. Cigarette smoking impairs the therapeutic response to oral corticosteroids in chronic asthma. *Amer J Respir Crit Care Med* 2003;**168**:1308–11.

62. Polosa R, Caponnetto P, Sands MF. Caring for the smoking asthmatic patient. *J Allergy Clin Immunol* 2012;**130**(5):1221–4.

63. Polosa R, Morjaria J, Caponnetto P, Caruso M, Strano S, Battaglia E, et al. Effect of smoking abstinence and reduction in asthmatic smokers switching to electronic cigarettes: evidence for harm reversal. *Int J Environ Res Public Health* 2014;**11**(5):4965–77.

64. Polosa R, Morjaria JB, Caponnetto P, Caruso M, Campagna D, Amaradio MD, et al. Persisting long term benefits of smoking abstinence and reduction in asthmatic smokers who have switched to electronic cigarettes. *Discov Med* 2016;**21**(114):99–108.

65. MacNee W. Pathogenesis of chronic obstructive pulmonary disease. *Proc Am Thorac Soc* 2005;**2**(4):258–66, discussion 90–1.

66. Morjaria JB, Malerba M, Polosa R. Biologic and pharmacologic therapies in clinical development for the inflammatory response in COPD. *Drug Discov Today* 2010;**15**(9–10): 396–405.

67. Fletcher C, Peto R. The natural history of chronic airflow obstruction. *Br Med J* 1977;**1**(6077):1645–8.

68. Hersh CP, DeMeo DL, Al-Ansari E, et al. Predictors of survival in severe, early onset COPD. *Chest* 2004;**126**:1443–51.

69. Anthonisen NR, Connett JE, Kiley JP, Altose MD, Bailey WC, Buist AS, et al. Effects of smoking intervention and the use of an inhaled anticholinergic bronchodilator on the rate of decline of FEV1. The lung health study. *JAMA* 1994;**272**(19):1497–505.

70. Burchfiel CM, Marcus EB, Curb JD, Maclean CJ, Vollmer WM, Johnson LR, et al. Effects of smoking and smoking cessation on longitudinal decline in pulmonary function. *Am J Respir Crit Care Med* 1995;**151**(6):1778–85.

71. Kanner RE, Connett JE, Williams DE, et al. Effects of randomized assignment to a smoking cessation intervention and changes in smoking habits on respiratory symptoms in smokers with early chronic obstructive pulmonary disease: the lung health study. *Am J Med* 1999;**106**:410–16.

72. Scanlon PD, Connett JE, Waller LA, Altose MD, Bailey WC, Buist AS, et al. Smoking cessation and lung function in mild-to-moderate chronic obstructive pulmonary disease. The lung health study. *Am J Respir Crit Care Med* 2000;**161**:381–90.

73. Tashkin DP, Rennard S, Taylor Hays J, Lawrence D, Marton JP, Lee TC. Lung function and respiratory symptoms in a 1-year randomized smoking cessation trial of varenicline in COPD patients. *Respir Med* 2011;**105**:1682–90.

74. Feldman C, Anderson R. Cigarette smoking and mechanisms of susceptibility to infections of the respiratory tract and other organ systems. *J Infect* 2013;**67**(3):169–84.

75. Sopori M. Effects of cigarette smoke on the immune system. *Nat Rev Immunol* 2002;**2**(5):372–7.

76. Wark PA, Johnston SL, Bucchieri F, Powell R, Puddicombe S, Laza-Stanca V, et al. Asthmatic bronchial epithelial cells have a deficient innate immune response to infection with rhinovirus. *J Exp Med* 2005;**201**:937–47.

77. Campagna D, Amaradio MD, Sands MF, Polosa R. Respiratory infections and pneumonia: potential benefits of switching from smoking to vaping. *Pneumonia* 2016;**8**:4. Available from: http://dx.doi.org/10.1186/s41479-016-0001-2.

78. Primatesta P, Falaschetti E, Gupta S, et al. Association between smoking and blood pressure: evidence from the health survey for England. *Hypertension* 2001;**37**(2):187–93.

79. Al-Safi SA. Does smoking affect blood pressure and heart rate? *Eur J Cardiovasc Nurs* 2005;**4**(4):286–9.

80. Fagard RH. Smoking amplifies cardiovascular risk in patients with hypertension and diabetes. *Diabetes Care* 2009;**32**(Suppl 2):S429–31.

81. Zanchetti A, Hansson L, Dahlöf B, et al. Effects of individual risk factors on the incidence of cardiovascular events in the treated hypertensive patients of the Hypertension Optimal Treatment Study. HOT Study Group. *J Hypertens* 2001;**19**(6):1149–59.

82. Mancia G, Fagard R, Narkiewicz K, et al. 2013 ESH/ESC Guidelines for the management of arterial hypertension. The task force for the management of arterial hypertension of the European Society of Hypertension (ESH) and of the European Society of Cardiology (ESC). *Eur Heart J* 2013;**34**:2159–219.

83. Virdis A, Giannarelli C, Fritsch Neves M, et al. Cigarette smoking and hypertension. *Curr Pharm Des* 2010;**16**(23):2518–25.

84. Farsalinos K, Cibella F, Caponnetto P, Campagna D, Morjaria JB, Battaglia E, et al. Effect of continuous smoking reduction and abstinence on blood pressure and heart rate in smokers switching to electronic cigarettes. *Intern Emerg Med* 2016;**11**(1):85–94.

85. Vasan RS, Larson MG, Leip EP, et al. Assessment of frequency of progression to hypertension in non-hypertensive participants in the Framingham Heart Study: a cohort study. *Lancet* 2001;**358**:1682–6.

86. Qureshi AI, Suri MK, Kirmani JF, et al. Is prehypertension a risk factor for cardiovascular diseases? *Stroke* 2005;**36**:1859–63.

87. Prescott E, Scharling H, Osler M, Schnohr P. Importance of light smoking and inhalation habits on risk of myocardial infarction and all cause mortality. A 22 year follow up of 12149 men and women in the copenhagen city heart study. *J Epidemiol Community Health* 2002;**56**:702–6.

88. Tonstad S, Cowan JL. C-reactive protein as a predictor of disease in smokers and former smokers: a review. *Int J Clin Pract* 2009;**63**:1634–41.

89. Doonan RJ, Hausvater A, Scallan C, Mikhailidis DP, Pilote L, Daskalopoulou SS. The effect of smoking on arterial stiffness. *Hypertens Res* 2010;**33**:398–410.

90. Niskanen L, Laaksonen DE, Nyyssonen K, Punnonen K, Valkonen VP, Fuentes R, et al. Inflammation, abdominal obesity, and smoking as predictors of hypertension. *Hypertension* 2004;**44**:859–65.

91. Hatsukami DK, Kotlyar M, Allen S, Jensen J, Li S, Le C, et al. Effects of cigarette reduction on cardiovascular risk factors and subjective measures. *Chest* 2005;**128**:2528–37.

92. Bolliger CT, Zellweger JP, Danielsson T, van Biljon X, Robidou A, Westin A, et al. Influence of long-term smoking reduction on health risk markers and quality of life. *Nicotine Tob Res* 2002;**4**:433–9.

93. Lee DH, Ha MH, Kim JR, Jacobs Jr. DR. Effects of smoking cessation on changes in blood pressure and incidence of hypertension: a 4-year follow-up study. *Hypertension* 2001;**37**:194–8.

94. Halimi JM, Giraudeau B, Vol S, Caces E, Nivet H, Tichet J. The risk of hypertension in men: direct and indirect effects of chronic smoking. *J Hypertens* 2002;**20**:187–93.

95. Perkins KA, Epstein LH, Marks BL, Stiller RL, Jacob RG. The effect of nicotine on energy expenditure during light physical activity. *N Engl J Med* 1989;**320**:898–903.

96. Poulter NR. Independent effects of smoking on risk of hypertension: small, if present. *J Hypertens* 2002;**20**:171−2.

97. Klesges RC, Meyers AW, Klesges LM, La Vasque ME. Smoking, body weight, and their effects on smoking behavior: a comprehensive review of the literature. *Psychol Bull* 1989;**106**(2):204−30.

98. Lycett D, Munafo M, Johnstone E, Murphy M, Aveyard P. Associations between weight change over 8 years and baseline body mass index in a cohort of continuing and quitting smokers. *Addiction* 2010;**106**:188−96. Available from: http://dx.doi.org/10.1111/j.1360-0443.2010.03136.x.

99. Eisenberg D, Quinn BC. Estimating the effect of smoking cessation on weight gain: an instrumental variable approach. *Health Serv Res* 2006;**41**:2255−66.

100. Russo C, Cibella F, Caponnetto P, Campagna D, Maglia M, Frazzetto E, et al. Evaluation of post cessation weight gain in a 1-year randomized smoking cessation trial of electronic cigarettes. *Sci Rep* 2016;**6**:18763. Available from: http://dx.doi.org/10.1038/srep18763.

101. Aubin H-J, Farley A, Aveyard P. Weight gain in smokers after quitting cigarettes: meta-analysis. *BMJ* 2012;**10**:345−439.

102. Audrain JE, Klesges RC, Klesges LM. Relationship between obesity and the metabolic effects of smoking in women. *Health Psychol* 1995;**14**:116−23.

103. Mokdad AH, et al. Prevalence of obesity, diabetes, and obesity-related health risk factors. *JAMA* 2001;**289**:76−9.

104. Wannamethee SG, Shaper AG, Perry IJ. Smoking as a modifiable risk factor for type 2 diabetes in middle-aged men. *Diabetes Care* 2001;**24**:1590e5.

105. Polosa R, Rodu B, Caponnetto P, Maglia M, Raciti C. A fresh look at tobacco harm reduction: the case for the electronic cigarette. *Harm Reduct J* 2013;**10**:19. Available from: http://dx.doi.org/10.1186/1477-7517-10-19.

106. Chinn S, et al. Smoking cessation, lung function, and weight gain: a follow-up study. *Lancet* 2005;**365**(9471):1629−35, discussion 1600−1.

107. Stewart ST, Cutler DM, Rosen AB. Forecasting the effects of obesity and smoking on U.S. life expectancy. *N Engl J Med* 2009;**361**:2252−60. Available from: http://dx.doi.org/10.1056/NEJMsa0900459.

INDEX

Note: Page numbers followed by "*f*" and "*t*" refer to figures and tables, respectively.

Printed in the United States
By Bookmasters